O9-CFU-483

LifebytheNumbers

Life

BY THE

Numbers

Keith Devlin

JOHN WILEY & SONS, INC.

NEW YORK • CHICHESTER • WEINHEIM • BRISBANE • SINGAPORE • TORONTO

This book is printed on acid-free paper. ♾

Published by John Wiley & Sons, Inc.
Published simultaneously in Canada

Design by Howard Grossman

Library of Congress Cataloging-in-Publication Data:

Devlin, Keith J.
Life by the numbers / Keith Devlin.
p. cm.
Includes index.
ISBN 0-471-24044-3 (cloth : alk. paper)
1. Mathematics—Popular works. 2. Life by the numbers (Television program) I. Title.
QA93.D458 1998
510—dc21 97-41059

Printed in the United States of America

10 9 8 7 6 5 4 3 2 1

Contents

Preface

Based on the television series by the same name, this is a book about everyday life and the role played in everyday life by mathematics.

It is not a "math book." It doesn't set out to show you how to do math. You won't learn much mathematics from this book, and you won't find any formulas or problems anywhere.

But you might well learn that mathematics is not at all what you thought it was. And you will definitely discover that there is hardly any aspect of your life in which mathematics does not play a significant—though generally hidden—part.

If you think that mathematics has little to do with your life, then this book is for you.

If you think that mathematics is just about numbers, then this book is for you.

If you think that mathematics was all worked out centuries ago, then this book is for you.

If you enjoyed the television series, then this book is for you. You will discover more about mathematics than it was possible to include in the series.

If you missed the television series, then this book is for you. Though based on the series, the book has been written to stand alone.

If you are curious about life—about sports, about entertainment, about art, about music, about gambling, about different kinds of professions, about computers, about animals, about deep sea exploration, about astronomy, about love and marriage, about . . . well, practically anything under (or indeed beyond) the sun—then this book is for you.

As a consultant on the television series, I was involved in some of the work that found its way onto the television screen, as were a number of other series consultants. But the main credit for the series goes to its producers, David Elisco, Joe Seamans, Gina Cantazarite, Mary Rawson, and Randy Quinn. They are the ones who did most of the work in developing the initial themes, carrying out the research, locating the appropriate film stock, and recording the many hours of on-camera interviews. By making available to me the rough-cut tapes of the programs and the transcripts of all the original interviews, they made my work as author of this book far easier than would otherwise have been the case.

As anyone who has seen the television series will know, the series producers did a marvelous job of bringing onto the screen a fascinating group of individuals, from all walks of life. Viewing the tapes and reading the transcripts, I decided that, in writing this book, I would try as much as possible to let those individuals speak for themselves—and for mathematics.

Of course, books and television are different media, so there are ways in which the book and the series differ. To make it possible for people to use the book to supplement the series or vice versa, I organized the book in chapters corresponding to the episodes of the series, using the same titles for my chapters that the producers did for the episodes of the series. I added an introductory chapter to set the scene for the rest of the book and a brief concluding chapter. And I changed the order of the chapters a little from the order in which the series was broadcast to provide better continuity in book form. I was also able to bring out larger themes and connections between different topics than was possible in a television series. But for all that, it remains "the book of the series."

While I was consulting on the television series, I was completing my book *Goodbye, Descartes* for John Wiley & Sons. My editor on that book, Emily Loose, was eager all along for the two of us to work together on a new book to accompany the television series. For my part, I had found working with Emily such a positive experience that I was as eager as she to try to secure the book contract for *Life by the Numbers,* and work together a second time. I am delighted that she was successful, and I hope that our delight shines through in the pages of this book.

Keith Devlin

MORAGA, CALIFORNIA
OCTOBER 1997

Chapter 1

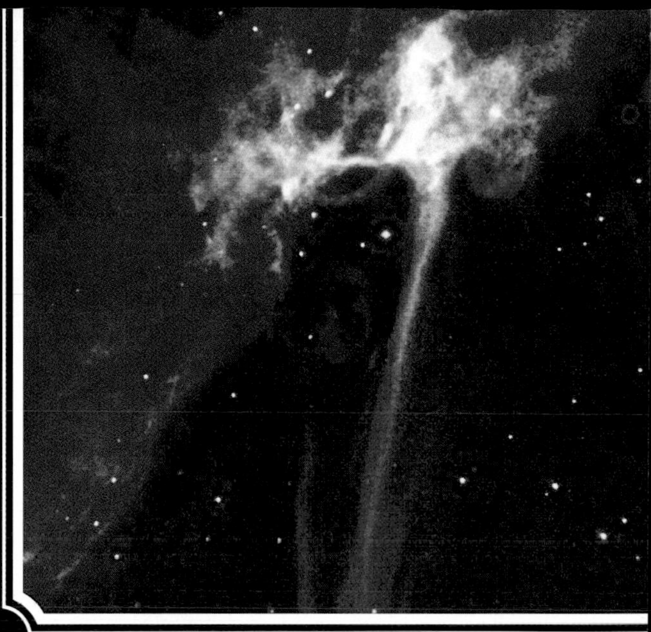

THE INVISIBLE UNIVERSE

For many people the mere mention of the word *mathematics* conjures up memories of complicated rules and dry arithmetic drills. But the truth is that mathematics as it is practiced by a remarkable range of people—from undersea explorers to special-effects designers—is creative,

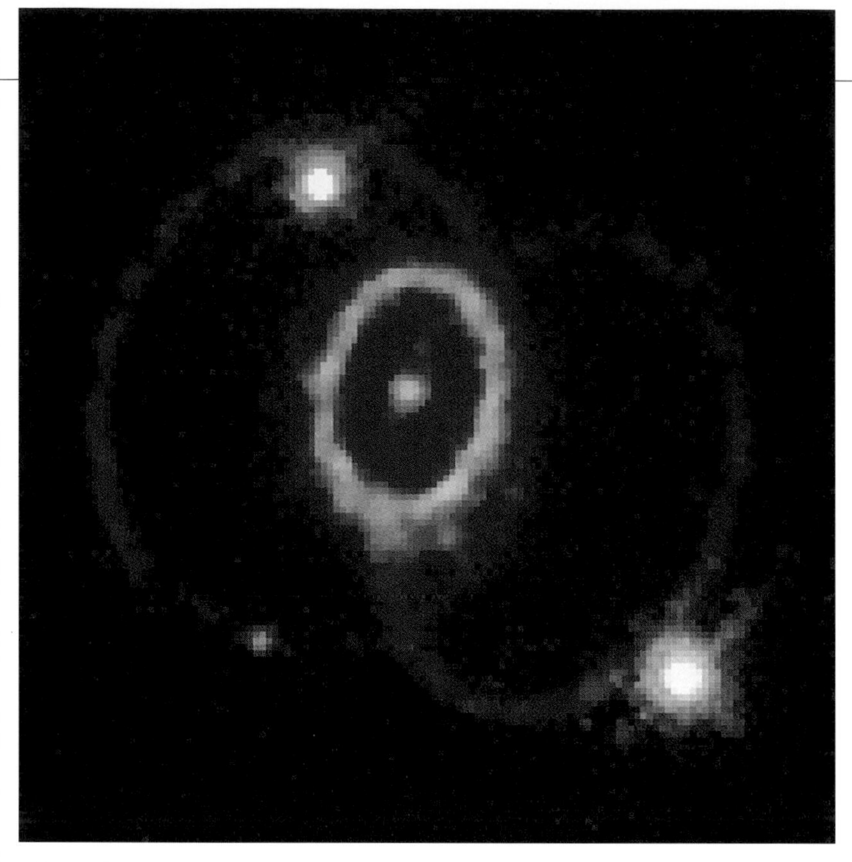

The patterns of mathematics are found all around us, from the smallest particles to the farthest reaches of the universe, as in the symmetrical rings of this supernova explosion.

The rules and the procedures learned in school are merely the *tools* you need to do "real" mathematics. Mathematics—real mathematics—is about trying to understand ourselves and the world we live in. Mathematicians take their inspiration from a surprising range of sources—questions about the origin of the universe, sports, or even children's stories. They use mathematics to investigate things that the eye cannot see, from the ocean deeps to the interiors of the stars. They develop methods to help in the fight against killer viruses, and they let us look inside the human mind. They use mathematics to map out our world and the cosmos, to help us to understand how trees and flowers grow, and to create new worlds—for entertainment and for exploration.

This book is not about the often dull and tedious mechanical aspects of mathematics. It's about the exciting things that can be done with mathematics. It's about things that we take for granted that would never exist without mathematics. It's about life. It's about trying to find answers to questions so simple that only a child would ask them.

HOW DOES THE LEOPARD GET ITS SPOTS?

For James Murray, it all began in the 1960s, when he was reading his daughter a bedtime story: "How the Leopard Got Its Spots," by Rudyard Kipling. In the story, an Ethiopian tribesman touches five fingers, drawn close together, all over the back of a leopard, and wherever his five fingers touch they leave five little spots in a cluster. Forever after, this beautiful arrangement of spots became the leopard's distinctive marking.

Murray's daughter loved the story. There was just one thing she wanted to know: "How does the leopard *really* get its spots?" Murray did not have the answer, but he told her he would find out. As a mathematician at the University of Oxford in England, he knew plenty of top-class biologists. He would ask one of them.

"When I'm walking in the woods, I find it quite difficult not to look at a fern or the bark of a tree and wonder how it was formed—why is it like that?"
JAMES MURRAY
mathematician

The "five fingers" marking of the leopard can be seen on the back of this leopard lounging in a tree.

"Great coaching and good intentions are not worth anything unless the goal is correct. The mathematical analysis is like turning on a light in a dark room."

KATHY CASEY
U.S. Olympic figure skating coach

He did just that. But to his surprise, none of them could answer his question. They knew that any coloration of an animal's coat is caused by a chemical called melanin, which is produced by cells just beneath the surface of the skin. It's the same chemical that makes fair-skinned people develop a tan when they are exposed to the sun. But why spots? Science did not have an explanation. As Murray discovered, no one knew how the leopard gets its spots. Or how the tiger gets its stripes. Or the zebra.

His curiosity aroused, Murray decided to try to find out for himself. It took him over twenty years. Today he has drafted his own, scientific version of Kipling's bedtime story, written in the language of mathematics.

HOW DO SKATERS PERFORM A TRIPLE AXEL?

Shelby Lyons and Damon Allen are two young skaters who share the same dream: they both want to win gold medals in the Olympic Games. Working with them at the Olympic Training Center in Colorado Springs, coach Kathy Casey is trying to help them achieve their dream. To do that, she has to figure out how to make the 200 bones, 600 muscles, and almost 100 joints of the human body work together to defy gravity and create airborne grace.

On a television monitor, Casey watches Damon perform a nearly faultless triple axel—three complete body rotations in midair. Once considered a daredevil maneuver, the triple axel was introduced into regular international competition by the Eastern Europeans in the 1980s. The entire maneuver lasts less than a second, but in that second lies the difference between winning a medal and going home empty-handed. Today, no skater can hope to win a major competition without a perfectly executed triple axel. As Casey says, "What it boils down to is, if you can't do the triple axel, you're toast."

Shelby Lyons performing a double axel.

When the triple axel first came onto the international scene, even the best American skaters were knocked out of the medal positions. To respond to the challenge, Casey turned to the new science of biomechanics. "Analyze the triple axel and tell me how to teach my skaters to do it," she asked the scientists. Among the basic questions Casey wanted to answer was, Is the secret to jump higher or to spin faster, or do you need some combination of the two?

In order to provide Casey with the information she needed, the entire maneuver had to be translated into the language of mathematics. For the mathematicians, finding the right answer was a purely scientific question. For the United States, it was a matter of national pride. For U.S. skaters like Shelby and Damon, it represented their only hope for a gold medal.

Today, as a result of the research Casey commissioned, U.S. skaters are once again the equal of any in the world. It's not that mathematics replaced the need for talent, skill, training, good coaching, and sheer determination. But mathematics provided the necessary direction.

HOW DID THE UNIVERSE BEGIN?

Mathematics has helped unlock the secrets of animal markings and figure skating techniques. It can also give us a way to look back to the origin of the universe, to try to understand the secrets of existence itself.

Several hundred never before seen galaxies are visible in this "deepest-ever" view of the universe, called the Hubble Deep Field, made with NASA's Hubble Space Telescope. The image reveals a bewildering variety of galaxy shapes, and some of these galaxies may be among the oldest in the universe.

In a small, windowless room in Champaign, Illinois, called the CAVE (CAVE Automatic Visualization Environment), artist Donna Cox is experiencing the birth of the universe. Wearing special stereovision spectacles, she watches as newly formed stars rush past her in a dramatic display of light and color.

The project Cox is working on began more than a year earlier with a mathematical model, a set of equations produced by the physicists. Those equations described what was going on in the universe in the first few seconds after the Big Bang—the cosmic explosion

that, according to science, was how our universe began. By feeding the mathematical model into a powerful computer, scientists obtained data—masses of data—that provided a step-by-step account of those first few moments of existence, when the ancestors of over 2 million stars were born.

The problem was how to comprehend that data. There was just too much of it for the human mind to grasp. As Cox says, "Data is a burden. We've got so much of it, and it's very much like taking twenty pounds of mashed potatoes and just shoving it through a straw."

This was why Cox was on the project in the first place. Trained as a graphic artist, for many years she had been collaborating with mathematicians and scientists at the University of Illinois in what was called a Renaissance Team. The scientists would bring data—long, long strings of numbers—to her, and she would work with the scientists to find ways to turn those numbers into pictures. By projecting those images onto the walls, ceiling, and floor of the CAVE and viewing them through stereovision spectacles, Cox and her scientist colleagues can travel through the data, experiencing the world that the numbers represent.

A question often asked of Cox is, How does her work differ from some sort of virtual-reality version of *Star Trek*? After all, both use computer graphics to create images. Cox has an answer: "With computer graphics, there are certain areas in advertising and entertainment where the goal is to create an illusion. The goal for my work with scientists and mathematicians is not to create an illusion but to reveal what's in the numbers. And that's a very different goal." For Cox, the aim is not to create an artificial world; instead, she wants to help us use our senses to understand the real world.

When we use mathematics to look back in time to the very origins of the universe, we make visible what would otherwise be invisible.

These three images translate complex scientific data into a striking visualization of two galaxies colliding.

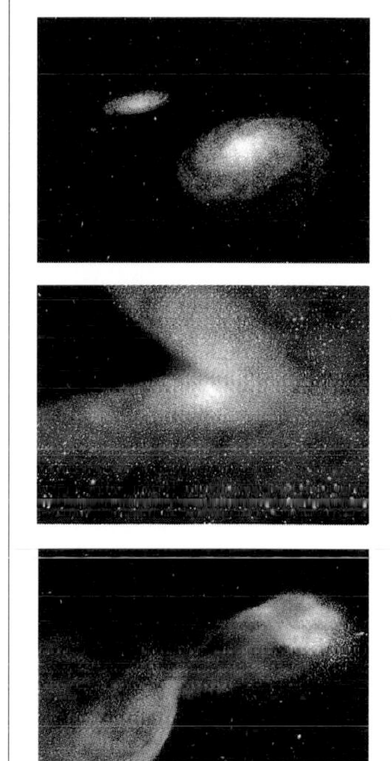

We can also use mathematics to see another otherwise invisible world: the world of the ocean floor.

IS THE OCEAN FLOOR FLAT?

On a ship in the South Pacific, geographer Dawn Wright examines television pictures of a brightly colored mountainous terrain. The images she sees are being created from the echoes of sound waves beamed from underneath the ship's hull toward the seabed, three miles beneath the ocean surface. Wright, who hails from Oregon, is a modern-day explorer. The unknown territory she is mapping out is the mysterious terrain at the bottom of the sea. "People used to think that the sea floor was flat and barren," Wright observes, "but what we're finding now is that the topography can be very, very rugged, very exciting."

Using mathematics to draw maps of the seabed, Wright is aware that her work is very different from that of the pioneers who first charted the North American continent. As she says, "I'm making maps of places that people have never been to and probably will never be able to go to."

Wright needs mathematics for her work to make up for not being able to see the ocean floor firsthand; she has to reconstruct images of the seabed on a computer from the data obtained by bouncing sound waves from the ocean floor. But for all her dependence on mathematics, Wright admits it is not her favorite subject. She leaves the mathematical part to others. "Mathematics is extremely important for making accurate maps of the sea floor," she admits, before going on to say, "The good thing about it is that you don't need to be a hard-core mathematician to do it."

Why does she do it? What is it about science that excites her so much? Wright provides the answer by recalling what got her into

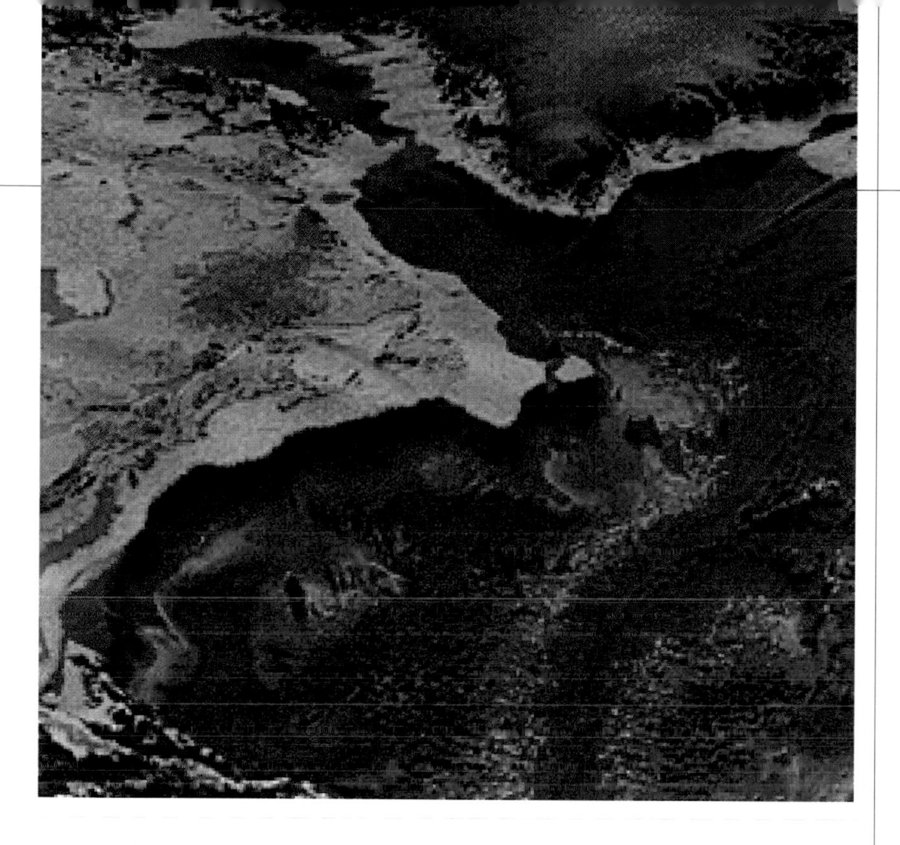

Dawn Wright produced this computer-generated image of the exotic terrain of the ocean floor using data from her explorations.

science in the first place. "The most important thing for kids who want to explore the earth nowadays is they have to really love the earth and be excited about it. Jacques Cousteau said that people protect what they love. And I'll add to that that people protect what they understand as well."

Dawn Wright sees it as important for people to understand our world. When we understand, we will protect, she says. Understanding is also the first step in taking action: action to protect the environment, or action to fight deadly enemies such as killer viruses. Mathematics can help there as well.

HOW CAN YOU RECOGNIZE A VIRUS?

Sylvia Spengler is a biologist and De Witt Sumners is a mathematician. Together they are engaged in a fierce battle against an enemy they cannot see: viruses. By understanding the way viruses work, Spengler and Sumners hope to provide clues for ways to overcome them.

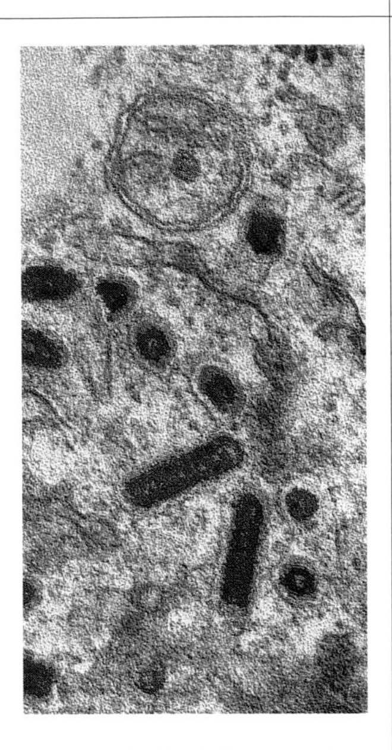

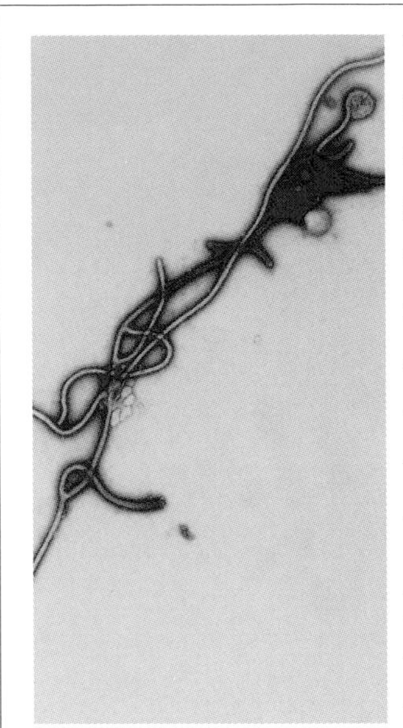

On the left, an electron microscope photograph of the rabies virus, and on the right, the knotty shape of the Ebola virus.

Viruses are one of the oldest, and simplest, forms of life on Earth, and also one of the most efficient. Viruses multiply and survive by getting others to work for them. When a virus lands on a cell, it injects itself into the cell and takes control of the cell. Once in command, it instructs the cell to make more copies of the virus until the cell explodes, spreading the virus to neighboring cells.

For some viruses, such as the common cold virus, the human body has developed ways to prevent the virus from spreading too far, so the victim recovers. But for other viruses, such as HIV (the AIDS virus), the body does not have an effective countermeasure, and as a result the virus ultimately wins the battle and the victim grows ill and eventually dies.

One of the difficulties facing scientists who investigate viruses is that viruses are far too small to be seen. Just how does the virus manage to get into the cell? This is where Sumners comes into the picture. His area of specialization in mathematics is called "knot theory."

Knot theory began in the middle of the nineteenth century. As the name suggests, it is the study of knots. Two of the basic questions asked by knot theorists are, How can you describe a particular knot? and How can you be sure that two knots placed in front of you are really different?

The first question is about notation. Mathematicians use ordinary algebra to describe the patterns of arithmetic, and musicians use musical notation to describe the patterns of music. What notation should knot theorists use to describe the patterns of knots?

The reason the second question arises is that two knots that might appear to be different might turn out to be essentially the same. Anyone who has tried to unravel a garden hose knows this. Assuming that the hose was neatly coiled in the first place, strictly speaking it won't be knotted at all. But as we all know, most garden hoses have an infuriating habit of coiling around themselves in such a complicated way that the "unknottedness" is not at all apparent; they look knotted!

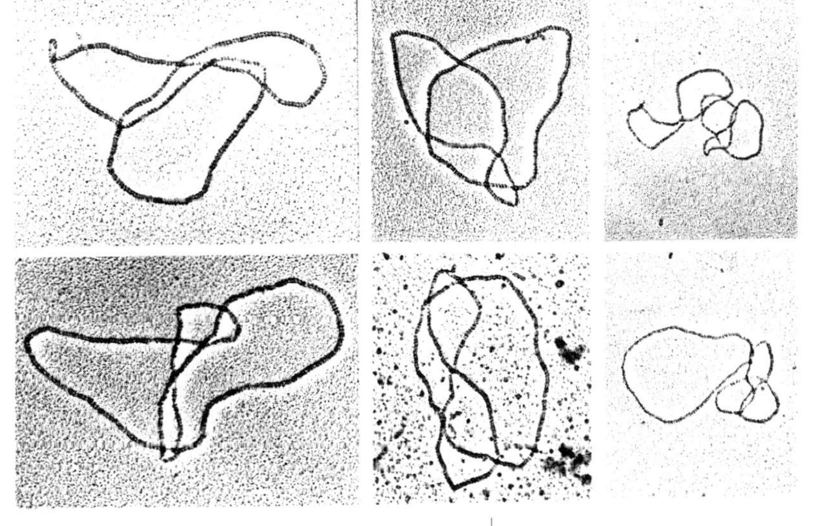

Electron microscope photographs of a selection of DNA knots.

While knots are of obvious interest to young scouts and to sailors, what on earth is the connection between knot theory and viruses that brings Spengler and Sumners together? It's all a question of nature once again trying to be efficient. A DNA molecule is like a long, thin string that coils up in order to fit inside the cell it occupies. In order to change the DNA to meet its own ends, a virus first manipulates the cell's DNA, tying it into a knot to bring certain parts closer together, and then swaps parts around and alters them. The knot pattern of the DNA produced by the virus in this process provides scientists with a kind of fingerprint to help identify the virus and understand the way it works, similar to the way a human fingerprint can help detectives identify a criminal and understand the way the crime was committed. As a result, knowing about knots is as important to "virus detectives" like Spengler, who want to investigate how a virus has broken into a cell, as fingerprints are to a police detective trying to find out who broke into a bank and how it was done.

IT'S NOT JUST NUMBERS

The importance of knot theory in the fight against viruses is relatively recent—the last twenty years or so. Prior to then, knot theory

The symmetries, stripes, and spirals of mathematical patterns can be found almost everywhere we look in the natural world.

was pursued by a small number of mathematicians, purely for curiosity's sake. This is typical. Time and again throughout history, a branch of mathematics that was developed with no particular application in mind has turned out to be extremely important to everyday life. The trick is to recognize when and where the mathematics comes in. Like geographer Dawn Wright, who does not need to "do" the mathematics in order to use it in her exploration of the seabed, biologist Sylvia Spengler did not have to become a mathematician in order to study viruses—she leaves that part of the work to her colleague, De Witt Sumners. What makes Wright and Spengler so effective in their work is that each was able to recognize when she needed mathematics, and how it could help her in her research.

Just under forty years ago, the scientist Eugene Wigner wrote an article titled "The Unreasonable Effectiveness of Mathematics in the Natural Sciences." "Why," asked Wigner, "is it the case that mathematics can so often be applied, and to such great effect?"

"It is true that a mathematician who is not somewhat of a poet will never be a perfect mathematician."
KARL WEIERSTRASS
mathematician (1902)

To appreciate Wigner's question, it's important to understand what he—and all other scientists—mean by "mathematics." To most nonscientists, mathematics is counting and calculating with numbers. That is not at all what a scientist means by the word. To a scientist, counting and calculating are part of arithmetic, and arithmetic is just one very, very small part of mathematics. Mathematics, the scientist says, is about order, about patterns and structure, and about logical relationships.

The patterns and relationships studied by the mathematician can be found everywhere in nature: the symmetrical patterns of flowers, the often complicated patterns of knots, the orbits swept out by planets as they move through the heavens, the patterns of spots on a leopard's skin, the voting pattern of a population, the relationship between the words that make up a sentence, the patterns of sound that we recognize as music—the list goes on. Sometimes the patterns are numerical and can be described and studied using arithmetic—voting patterns, for example. But often they are not numerical—the

"I think mathematics is very imaginative. I think what defines the people who are good at it and bad at it is whether they can see things in a totally different way. That takes a certain imagination."

JAMES MURRAY
mathematician

patterns of knots and the symmetrical patterns of flowers, for example, have little to do with numbers.

As the "science of patterns," mathematics is a way of thinking about the world—one way among many. Thinking mathematically about our world helps us to understand it. To take a simple example, what does it mean to say you "know" about flowers? To a gardener, knowing about a particular kind of flower means knowing about the conditions under which it grows best—what is the best soil, the right amount of water and nutrients, the ideal temperature, and so forth. To a florist, knowing about a flower means appreciating which other flowers it can be combined with to create an attractive bouquet, and knowing the conditions under which it preserves its appearance. To a botanist, knowing about a flower means understanding the functions of the various parts of the flower—the stem, the leaf, the petals, the stamens, and so on. A chemist would understand the flower in terms of the different chemical reactions that take place in the flower as it grows, takes in nutrients, and absorbs the rays of the sun. A mathematician can "know about" a particular flower in yet another way, in terms of its pattern of symmetry: how many different ways can you turn it so that it still "looks the same"? No one way of knowing about a particular flower is the "right" way. It all depends on what you want to do with your knowledge.

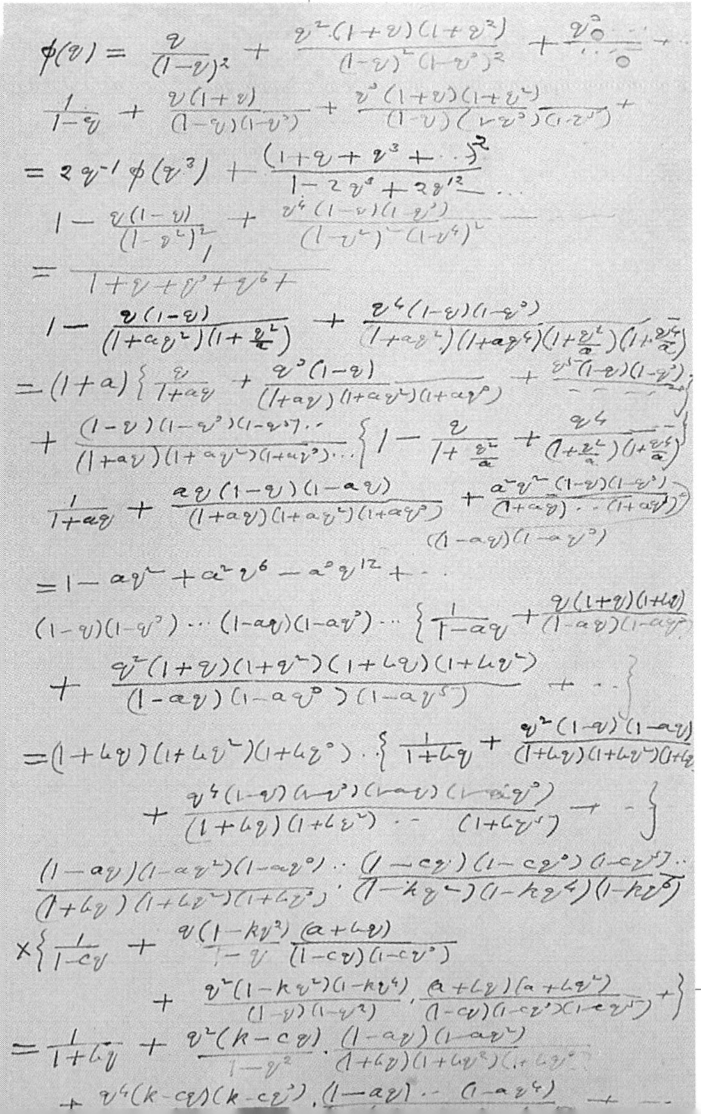

Once you realize that mathematics is not some game that people simply make up, but is about the patterns that arise in the world around us, Wigner's observation about the "unreasonable effectiveness of mathematics" does not seem quite so surprising. Mathematics is not about numbers. It is about life. It is about the world in which we live. It is about ideas. And far from being dull and sterile, as it is so often portrayed, it is full of creativity.

Jaron Lanier, who is both an accomplished concert pianist and a well-known computer scientist, puts it this way: "Humans didn't evolve looking at numerals. But humans did evolve doing mathematics, because every time you run, every time you jump, every time you catch a ball, you're doing mathematics—a very physical kind of mathematics. We have this enormous wisdom about us as we perform tasks, move with our bodies, move around in space.

"I'm absolutely convinced that mathematics is the most naturally human thing to be interested in, but it's taught with a language that's alien to many people. Even for people who are into mathematics, it's hard to get into the language—it's very dry, very technical."

The notations of mathematics can be considered comparable to musical notations. Opposite, a page from the "notebooks" of the famous Indian mathematician Srinivasa Ramanujan, who invented his own idiosyncratic style of mathematical notation. Left, page one of Ludwig von Beethoven's Violin Sonata no. 10 in G major, op. 96.

"Mathematics, rightly viewed, possesses not only truth, but supreme beauty—a beauty cold and austere, like that of sculpture . . . sublimely pure, and capable of a stern perfection such as only the greatest art can show."

BERTRAND RUSSELL
philosopher (1910)

For Lanier, it is no surprise that mathematics and music are two of his greatest passions: "There's a kind of beauty that exists in mathematics and in music that is so profound that if you love one, then you have to love the other. I don't know how I could say it better than that."

Obviously, Lanier's comparison of mathematics with music would make no sense if mathematics were simply arithmetic. Sadly, for most people, mathematics *is* no more than arithmetic. Most of mathematics remains hidden from view, its ubiquitous presence and its importance to our lives known only to a few.

As should become clear when you leaf through the pages of this book, not only is mathematics a rich and often beautiful product of the human mind—a major part of our culture—but there is almost no area of life that is not affected by it in deep and profound ways. Quite simply, mathematics is the "invisible universe."

SYMBOLS OF SUCCESS

The first thing that strikes anyone who opens a typical book on mathematics (not this book!) is that it is full of symbols—page after page of what, to most people, is utter gobbledygook. It's like a foreign language written in a strange alphabet. In fact, that's exactly what it is: a foreign language. Mathematicians express their ideas in the language of mathematics.

Why? If mathematics is about life and the world we live in, why do mathematicians use a language that turns many people off the subject before they are out of high school? It's not because mathematicians are perverse. Nor is it because they are a strange breed of individuals who like to spend their day swimming in an algebraic sea of meaningless symbols. The reason for the reliance on abstract symbols is that the patterns studied by the mathematician are abstract patterns.

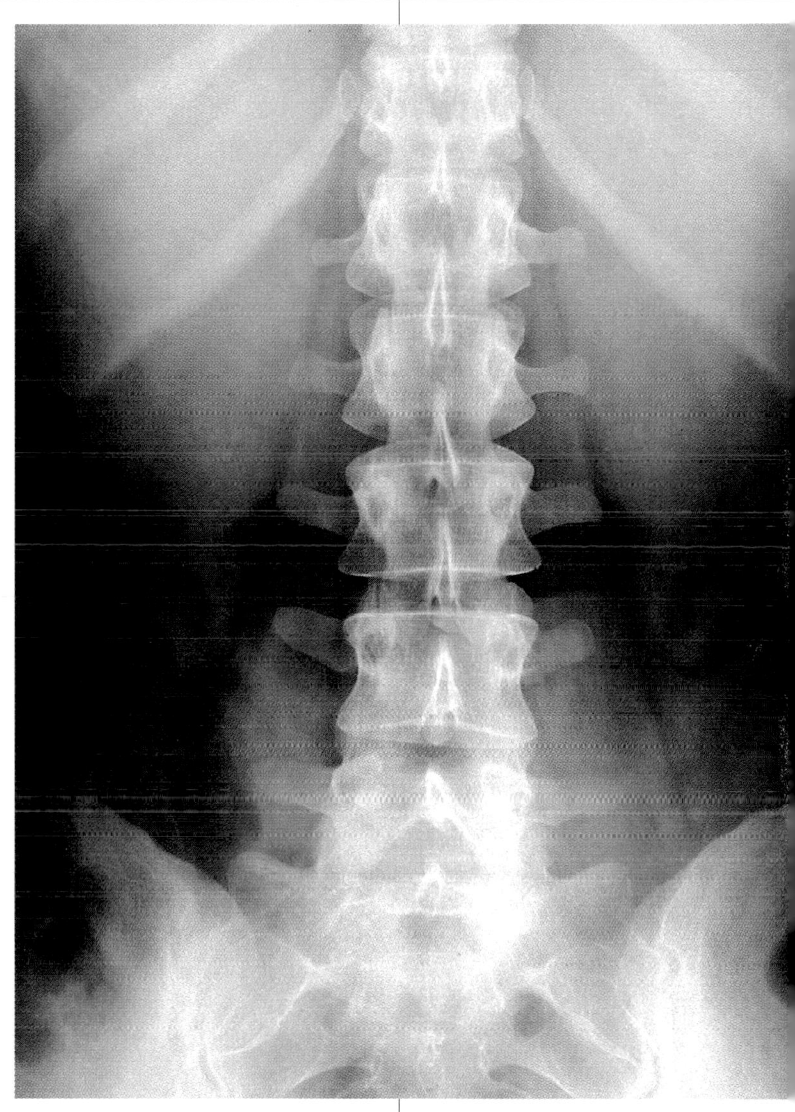

You can think of the abstract patterns of the mathematician as "skeletons" of things in the world. The mathematician takes some aspect of the world—say, a virus, a flower, or a game of poker—picks some particular feature of that virus, flower, or game, and then discards all the particulars, leaving just an abstract skeleton. In the case of the virus, the abstract pattern that is left—the skeleton—might be the knot pattern: how the DNA molecule wraps around itself. For a flower, it might be the pattern of its symmetry. For a poker game, it might be the distribution of the cards or the pattern of betting.

To study such abstract patterns, the mathematician has to use an equally abstract notation. Music provides a useful analogy. Musicians use a notation just as abstract as algebra to describe the patterns of music. Why do they do this? Because they are trying to describe on paper a very abstract pattern that exists in the human mind when we listen to music. The same tune can be played on a piano, a guitar, an oboe, a flute, and so on. Each instrument produces a different sound, but the tune is the same. What determines the tune, and distinguishes it from some other tune, is not the instrument but the pattern of notes produced by that instrument. It is that abstract pattern that the musician

The intricate structure of the spine undergirds the upright posture of the human body. Analogous mathematical structures lie hidden within all of the structures in our universe.

captures on the page using musical notation, not the particular sound of a specific instrument. To capture an abstract pattern, the musician, like the mathematician, needs an abstract notation.

When a mathematician looks at a page of mathematical symbols, he or she does not really "see" the symbols, any more than a trained musician "sees" the musical notes on a sheet of music. The trained musician's eyes read straight "through" the symbols to the sounds the symbols represent, as the musician "hears" the music in his or her mind. Similarly, a trained mathematician reads straight "through" the symbols to the patterns the symbols represent.

The essence of the mathematician's approach, and the key to its incredible power and success, lies in the extreme simplicity and highly abstract nature of the patterns singled out for study. Mathematicians take a highly simplistic view of the world, avoiding the complexities of nature and of life. The mathematician views the world in terms of perfectly straight lines, perfect circles, geometrically precise triangles, squares, and rectangles, smooth, flat planes, instants in time, and the like. In the real world there are no perfect circles, no perfectly straight lines or smooth, flat planes, and, to all intents and purposes, no instants in time. If you don't believe me, take a look at a glass plate (a "smooth, flat plane") through a microscope. What looks to the naked eye like a flat plane will reveal itself to be far from smooth and flat. The mathematician's circles, lines, planes, and so forth are all fictions, figments of the imagination. They are idealizations of the "almost circles," the "fairly straight lines," the "smooth-and-flat-to-the-naked-eye planes," et cetera, that we find in the real world.

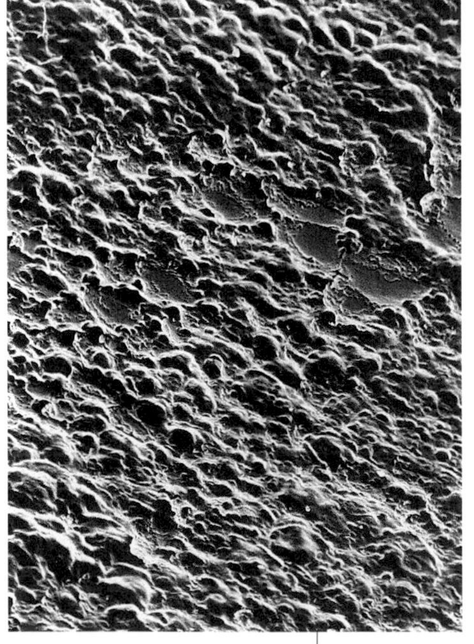

The surface of a sheet of glass is not really a flat plane. Here we see a glass surface magnified three hundred times.

A further analogy to mathematics and mathematical

notation is provided by blueprints for buildings and machinery. Blueprints are a kind of "skeleton" of the object they represent; they provide the basic structure, devoid of any of the complexities of the object.

A mathematician looking at the world and stripping away complexity to leave only the underlying structure or pattern can be likened to a physician taking an X ray of a patient—the X-ray machine strips away all images of the skin, flesh, and muscle, leaving only a picture of the underlying skeleton. That's another reason to call mathematics the invisible universe.

Once you realize what mathematics is really about—what mathematicians are really up to—it should not come as too much of a surprise to learn that mathematics lies behind—or beneath—most things in the world, and that there is scarcely any aspect of our lives in which mathematics does not have applications. In their ability to describe and reason about highly abstract patterns, the mathematician's symbols are without doubt both symbols of power and symbols of success.

This book will provide you with a glimpse of just a few of the many, many ways that mathematics—the science of patterns, the invisible universe—plays a role in our everyday life.

Chapter 2

SEEING IS BELIEVING

Doug Trumbull makes dreams come true—or rather, he makes it seem that

"Mathematics is a sort of hidden tool for me. In the world of computer graphics, mathematics is behind everything I do."
DOUG TRUMBULL
moviemaker

Angeles in *Blade Runner,* the alien spacecraft in *Close Encounters of the Third Kind,* and many others. These days, Trumbull spends a lot of his time developing a new kind of cinematic experience: immersion theater, where in addition to seeing a screen and hearing sound, the audience experiences motion.

Of his more recent work, Trumbull says, "My interest is in movies as an engulfing, immersive experience where the audience gets to actually enter into the movie and become part of the action. The barriers between you and what's on the screen are going away. We're giving the human senses, our ears, our eyes, our body, much more information, and that draws us into the movie like a real experience."

For Trumbull, the process begins with imagination—with an idea for a new ride, a new kind of experience. "My mind goes off into some kind of imaginary space," he explains. "I see images. . . . I see things that may or may not exist."

To translate that initial idea to a cinematic reality, Trumbull relies on mathematics. It is through mathematics that the idea becomes a blueprint—a specification—for a ride. Just as the mathematician can look at reality and see an underlying pattern—a skeleton or a blueprint—so too Trumbull can go in the opposite direction. He can start with the skeleton—the mathematical blueprint—and build a world on top of it. The resulting world will not be reality; it will be a virtual reality, a fake world. The trick is to make that fake world sufficiently like the real world for people to be able to experience it as real—to feel that they really are traveling through empty space or hurtling along a narrow canyon in a supersonic starfighter.

Harrison Ford hangs on for dear life in the simulated cityscape of *Blade Runner.*

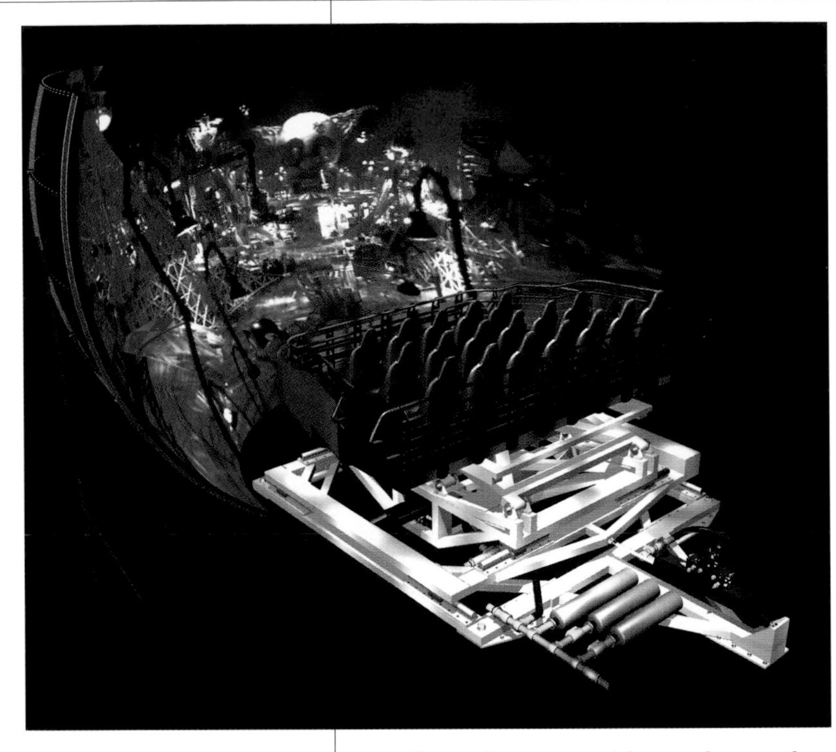

A fantastical carnival world comes alive on the screen of Doug Trumbull's simulator ride.

In one of Trumbull's more recent rides, for example, the seats move only thirty inches in any direction, but by having a screen that completely fills the audience's field of vision, and by closely linking the motion of the ride to the images shown on the screen, Trumbull creates the illusion of flying through space. Tying in the pictures with the motion in just the right way involves mathematics. Trumbull himself is no mathematician: "I have sort of a basic, intuitive understanding of geometry. I know that mathematics and numbers are behind everything I do." The actual mathematics he leaves to others.

Trumbull's approach is so common in making movies—using mathematics to create special effects—that the film industry is now one of the largest employers of mathematicians in the world. Movies such as *Terminator 2, Forrest Gump,* and *Star Wars* make particularly heavy use of the mathematician's talents to create images that look completely real but that are created entirely inside a computer. Movie companies such as Lucasfilms and Industrial Light and Magic, which specialize in special-effects sequences, are made up almost entirely of people highly skilled in mathematics.

In the movie *Forrest Gump,* there is a scene in which Tom Hanks, the star of the movie, shakes hands with President Kennedy, who had been dead for thirty years when the movie was made. How was this scene created? By combining old newsreel film of Kennedy shaking hands with some college students with a carefully filmed scene of Hanks shaking hands with . . . nobody. Both sequences were fed into the computer. What that means is that each frame from each film was digitized—turned into a mathematical representation as a massive array of numbers. The computer processed the numbers in such a way that when the final arrays—the answers to the calculations—were converted back into pictures, the result was a sequence in which we see Kennedy shaking hands not with some sixties student but with Hanks.

The special effects in movies such as *Terminator 2*, *Star Wars*, and *Jurassic Park* were similarly created. In all three cases, the spectacular pictures we see on the screen are the results of millions of arithmetic computations performed on a powerful computer. We see scenes and actions that were never performed on a movie set; they were created in a computer. Their only existence is within the world of mathematics.

Though you sometimes hear it said that "The world is full of mathematics," this is really just a figure of speech. Mathematics is not out there "in the world." Mathematics is inside our heads, a product of the human mind that has grown and developed over three thousand years. It is because mathematics gives us such a powerful way of looking at the world, and of seeing things in the world that would otherwise be invisible, that we are able to declare that "It's a mathematical world."

Mathematics provides us with a way to understand the world—to bring the world into our minds. In this respect, mathematics is like our scientific theories—physics, chemistry, and biology. But mathematics can

By designing his rides with a computer, Doug Trumbull can calculate the precise movements that will create the simulated experience of moving through the world on the screen.

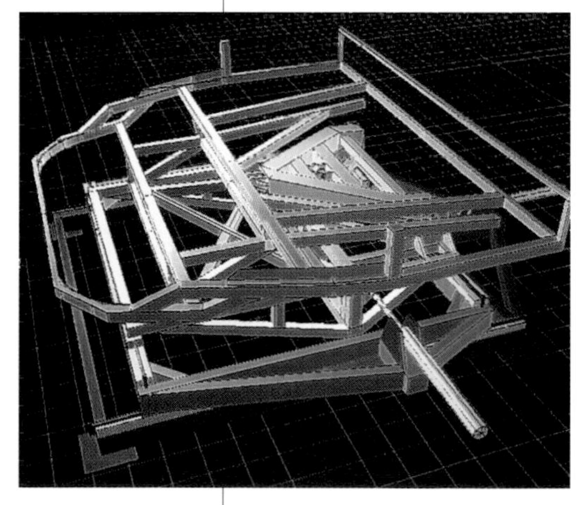

"The advent of linear perspective really mathematicized, or probably better, geometrized the way we see. The understanding of perspective allows a human being to make a picture of a controlled and confined space. It doesn't have to replicate a real space at all. And perhaps that's the real power of perspective: you're not simply copying nature, you're creating it."
SAM EDGERTON
artist and art teacher

also be used to take ideas that exist within our minds and put them outside in the world for others to share. Viewed in that way, mathematics is creative, like literature, painting, and music. We use it to take the ideas produced by our imagination and put them into plain view for others to see and experience—sometimes to inform, sometimes to stimulate or provoke, and other times to entertain.

These two ways of looking at mathematics—as an analytic science and as a creative art—cannot really be separated. They are just two sides of the same coin. Scientists often need to be creative and imaginative in using mathematics to understand the world, and the scientific aspects of mathematics can be necessary to express many inner ideas. Doug Trumbull makes use of the science of math to give expression to his creative visions, and in this respect he is one in a long line of creative artists who have crafted astonishing images from the application of mathematical rules.

AN OLD PERSPECTIVE

Trumbull sees his special-effects work as continuing a mathematical tradition that began with the painters of fifteenth-century Italy. It was, says Trumbull, the Renaissance artists who started the special-effects business, by figuring out how to make a two-dimensional image on a canvas appear to be three-dimensional—to look real, in other words.

What the Renaissance artists discovered, and what today's students learn in art school, is that to create the illusion of three dimensions on a two-dimensional canvas, you need to start with the right mathematical skeleton, and build your scene on top of that framework. The key was what we now call the mathematical theory of perspective.

In the painting opposite, "Scenes from the Bible" by Duccio Buoninsegna, painted in 1320, the stylized treatment of space characteristic of medieval painting is still used. The famous painting above, "The Annunciation" by Domenico Veniziano, painted in 1445, is one of the first examples of the use of vanishing point perspective.

The first idea behind perspective is that directly opposite the viewer's eye is an imaginary spot called the vanishing point or the "point at infinity." All lines that the viewer should see as parallel, going from the eye perpendicularly into the picture, should meet at the vanishing point. If the painting is drawn so that this is the case, it will appear to have depth. It might still look surreal and out of balance, however. To create true perspective, the artist also needs to arrange things so that lines that are meant to be parallel in any direction (not just perpendicular from the viewer's eyes) meet at an imaginary point on a single imaginary line, the so-called line at infinity. Any painting that faithfully obeys the constraints imposed by having supposedly parallel lines meet on the line at infinity will look three-dimensional.

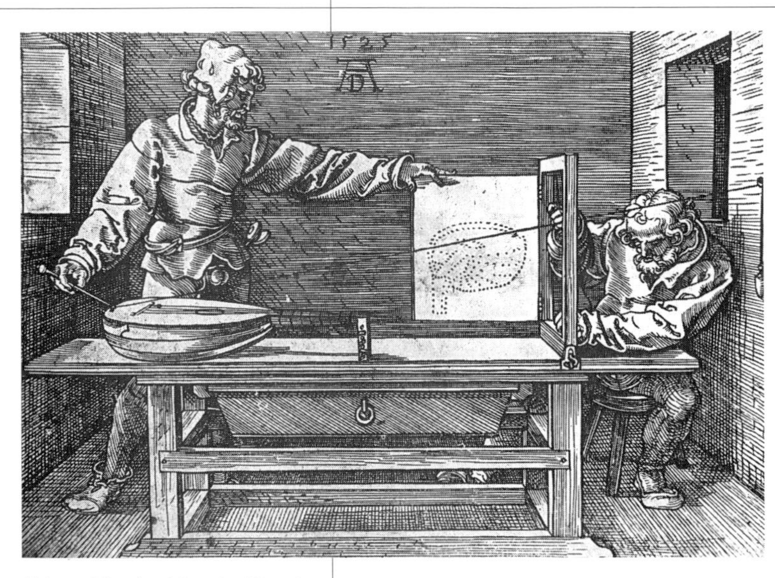

This etching by Albrecht Dürer is one of many he made depicting the methods by which artists of the time created the grid of lines that allowed them to introduce perspective in their works.

In Trumbull's own words, "When the Italians figured out perspective, paintings started to look three-dimensional rather than two-dimensional. For me, as an artist, I had this revelation when I was in art school and suddenly I understood perspective. It blew my mind to experience the same thing the Italians experienced then. That was revolutionary to me."

Today, we are so used to seeing "depth" in a two-dimensional image—in paintings, in photographs, on television, and in film—that it seems perfectly natural. But until Renaissance artists such as Leonardo da Vinci and Albrecht Dürer discovered the secrets of perspective drawing, it may well not have occurred to anyone that it was even possible to achieve such an effect. As artist and art instructor Sam Edgerton remarks, "Nobody is ever born to draw a picture in perspective, which is why it took so long to figure it out."

Each year, Edgerton teaches the geometry of perspective to a new class of art students. The key to painting, he tells them, is to realize that a painting is not the same as a photograph. A painter has to be aware of the way the human visual system interprets what it sees, and to use that knowledge to create visual illusions that the mind sees as "real." "You're following strict laws of geometry and you're creating a visual image of something you're not even seeing yet except in your mind's eye," he tells the class.

The geometry Edgerton is referring to is called projective geometry. It is the geometry of human visual perception, and as such is the geometry that lies behind perspective drawings. The basic principles of projective geometry were worked out in the years following the Renaissance artists' discovery of perspective, as mathematicians sought to understand the new secret the artists had discovered. The very first textbook on projective geometry was written in 1813. Its author was a young French mathematician called Jean-Victor Poncelet. A military officer, he wrote the book while a prisoner of war in Russia.

Projective geometry is one of two mathematical keys to modern computer graphics. By being programmed to obey the laws of projective geometry, a computer can produce an image that looks like a real scene or object, with depth and shadow. To do this, the computer programmer uses a technique that dates back to the seventeenth century and one of the most famous philosophers of all time: the Frenchman René Descartes. Best known for his famous remark "I think, therefore I am," Descartes made essentially just one contribution to mathematics. But what a contribution!

In 1637, Descartes published a book describing what is known today as the scientific method. In an appendix, he gives an account of a new way to do geometry. The idea is to reduce it to algebra, so that a problem in geometry is replaced by an equivalent problem in algebra. A present-day equivalent might be taking a successful movie and turning it into a novel. If the translation from screen to book is done well, it will be the same story in both cases, just described in different ways.

René Descartes, the philosopher and mathematician who so famously remarked, "I think, therefore I am."

In honor of Descartes, geometry done algebraically in the manner he described is generally referred to as Cartesian geometry.

There are two oft-repeated stories about Descartes's revolutionary new approach to geometry. Both could be true, though there is no hard evidence for either. One story is that the great philosopher was so bad at geometry that he was driven to find a way around it, and Cartesian geometry was the result. The other story is that Descartes, who was a frail individual prone to sickness, was lying ill in bed one day, when his eye was caught by a fly crawling around on the ceiling. Watching the fly move around, he wondered if he could describe its path by means of a numerical equation.

The rest, as they say, is history. Or rather, it was history. Today, it is something else, something as modern as you can get. As mentioned earlier, when Descartes's algebraic approach to geometry is combined with the projective geometry of a later generation, you get the basis of modern virtual-reality technology and the very latest movie special effects.

MATHEMATICS IN THE GALLERY

Donna Cox.

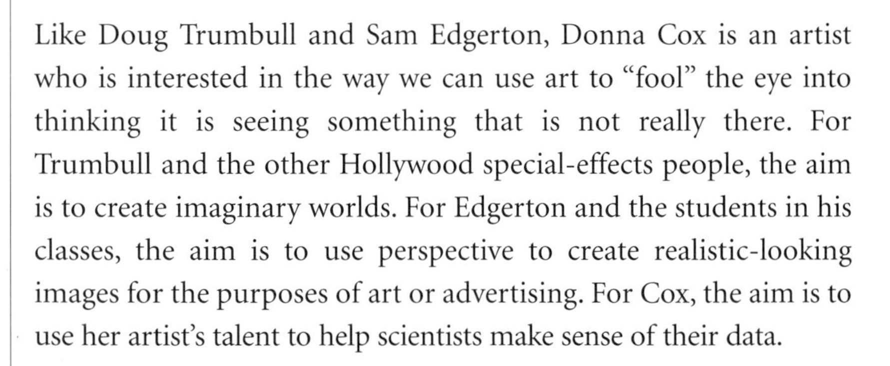

Like Doug Trumbull and Sam Edgerton, Donna Cox is an artist who is interested in the way we can use art to "fool" the eye into thinking it is seeing something that is not really there. For Trumbull and the other Hollywood special-effects people, the aim is to create imaginary worlds. For Edgerton and the students in his classes, the aim is to use perspective to create realistic-looking images for the purposes of art or advertising. For Cox, the aim is to use her artist's talent to help scientists make sense of their data.

Cox works at the University of Illinois at Urbana. Though a member of the art department, she is likely as not to be found over in the astronomy department, or talking to computer scientists and physicists in the National Center for Supercomputer Applications that is housed on the Urbana campus. "I was always torn between whether to major in art or in science," she recalls. After graduating in art, she finally resolved the dilemma by doing both. "Where I really found that things came together was with computer graphics," she says. These days, Cox's specialty, the reason why scientists continually flock to her to help them understand their data, is what is called scientific visualization.

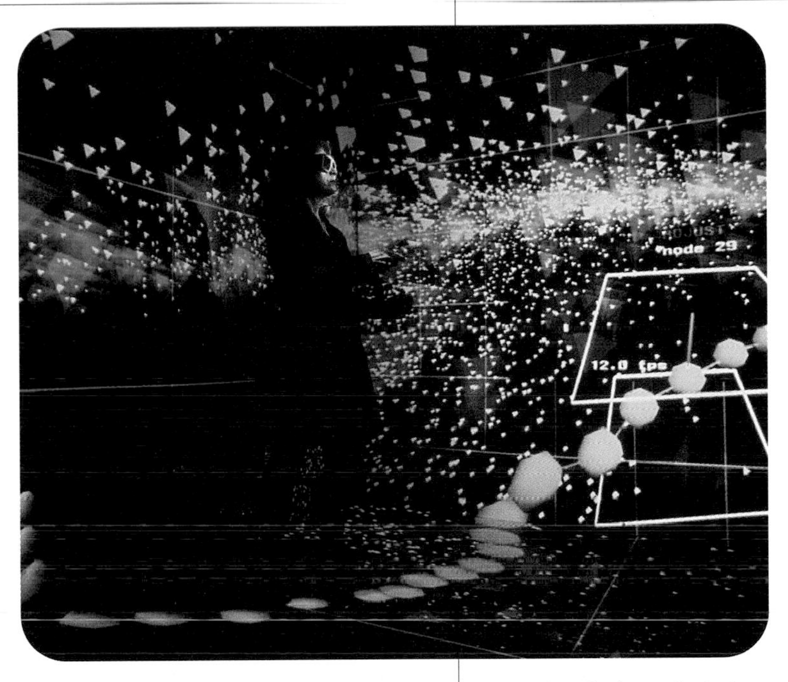

Donna Cox viewing a simulation of the first moments after the Big Bang in a virtual reality chamber.

As computers became more powerful, scientists were able to perform more and more calculations, creating more and more data. With the advent of computers that could generate a billion new numbers every second, the scientists soon found they were drowning in numbers. For the first time in history, they were able to generate lots of accurate data, but they could not understand it. The human mind simply cannot comprehend so many numbers. The data had to be converted into a form that humans could understand. That form was pictures—visual images. Cox uses her knowledge of art and human perception to find ways to represent data in a visual form that the human mind can comprehend. "We live in an image-glutted society," Cox observes. "With every advance of technology, we produce more and more information. I take information that has already been generated as numbers and use the best tools of our age to communicate it."

Cox uses the techniques and the technology the entertainment industry developed to fake reality in order to help scientists understand (real) reality. The results can sometimes be surprising. Occasionally, one of Cox's scientific visualizations turns out to have an unexpected beauty, a beauty in nature that had hitherto remained hidden from view. As Larry Smarr, director of the National Center for Supercomputer Applications, has observed,

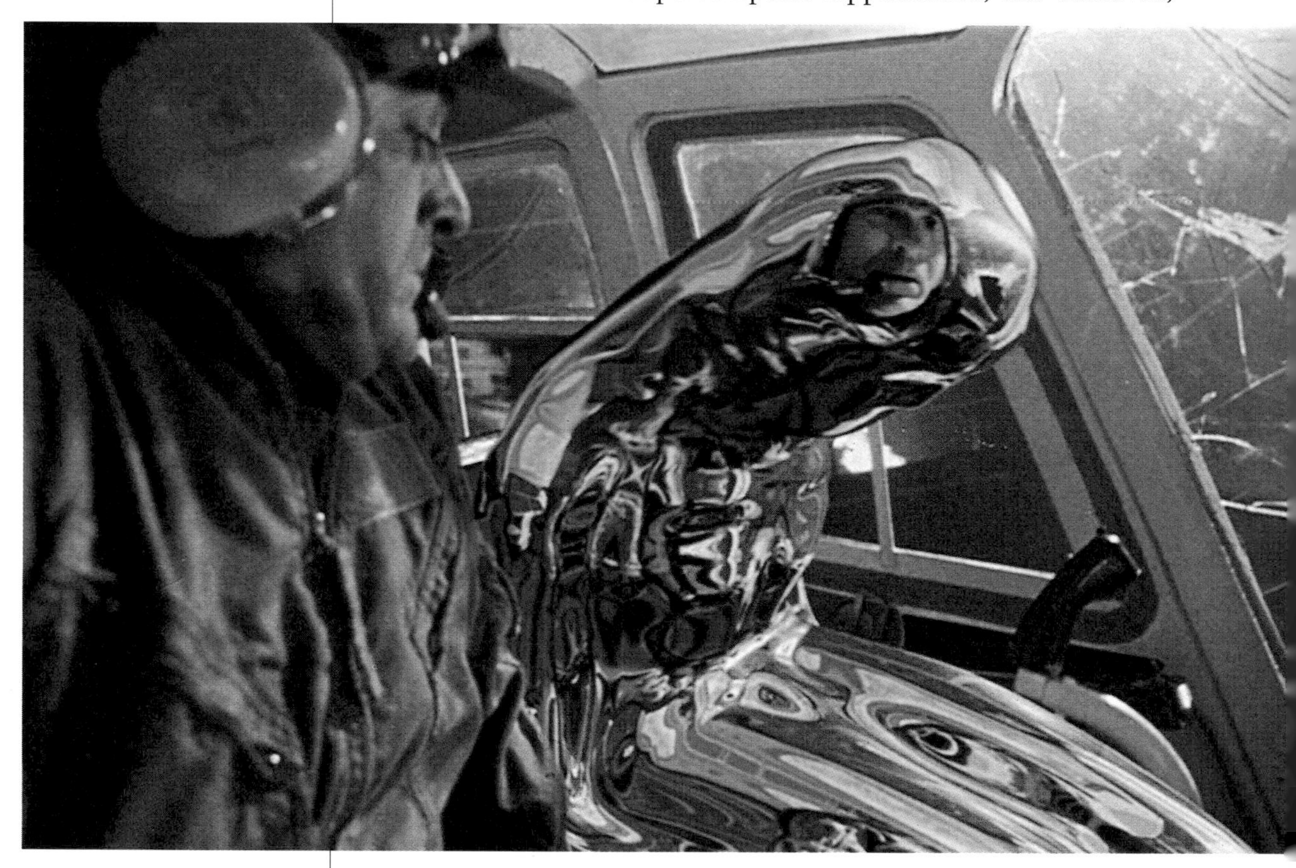

Morphing was put to fascinating use in the movie *Terminator 2,* in which the character of the terminator possessed the ability to dissolve at will into a metallic fluid form and to take on a vast range of sizes and shapes.

one of the unexpected things to come out of the collaboration between Cox and the scientists has been original art—art that can justifiably find its way onto the walls of the art gallery.

One particularly startling example of mathematical art came from the work Cox did with University of Illinois mathematician George

George Francis drawing a graphic representation of a complex topological surface.

Francis. Francis works in the area of mathematics known as topology. Topology studies what happens to figures when they are subjected to "smooth, continuous changes," where their shape alters but the figure is not torn or cut—as if the figure were made out of an indefinitely stretchable elastic material. (Topology is the branch of mathematics that lies behind those sequences in movies and television advertisements where a character or object having one shape is transformed in a fluid fashion into a quite different shape. The movie term for such a transformation is "morphing." The name comes from the mathematical term for such a transformation: a morphism.)

Unlike movie morphisms, mathematical morphisms can be extremely complicated. Most of the time, the mathematician has to rely on algebraic formulas to keep track of things. Occasionally, however, it is possible, though difficult, to draw a picture. Whenever this was possible, Francis would seize the opportunity. Though he could work with the symbols, he preferred the pictures. To help himself understand some of the intricacies of a particular morphism, he would draw elaborate pictures on the blackboard, using colored chalk. After a while, he got quite good at it. Seeing one of his drawings in his office one day, Donna Cox had an idea. "George, I do computer graphics; you do these wonderful surfaces. Why don't we collaborate and make these images with a supercomputer?"

They decided to give it a try. They chose two surfaces and set out to animate the process whereby one morphs into the other. Francis supplied the algebraic description for the morphism; Cox programmed the computer to display the morphing process on the screen. Neither of them had any idea what kinds of shapes they

"The best part [of collaborating with scientists and mathematicians] is to know that I can share information with another human being and that together we can make something that has never been created before. Making something that is invisible visible is one of the greatest joys that I have."

DONNA COX
artist

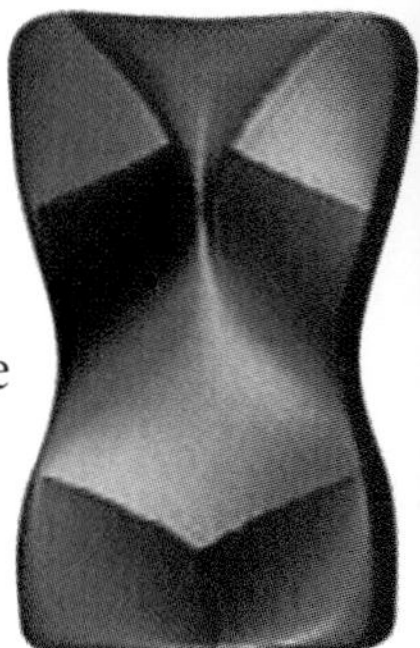

would see in between the start and finish surfaces. They expected to see the kinds of geometric figures you see in any mathematics textbook. Instead, as the morphing progressed, they suddenly found themselves looking at a classically proportioned female torso sitting in the middle of the computer screen. Remarking that the figure reminded them of a goddess, they decided to call it the Venus. It was one of the first pieces of "classical art" produced by a computer in the course of solving a mathematical problem. For both Francis and Cox, the discovery of the Venus confirmed that the abstract, logical beauty of mathematics and the visual beauty we see in the world around us can coincide.

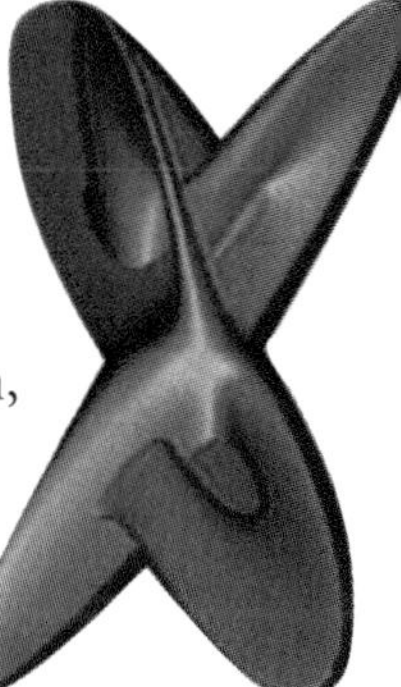

Though startled by the appearance of the female form, Francis was not surprised to see a beautiful shape on the screen. Any mathematician knows that the rather sterile-looking formulas of mathematics describe the underlying beauty that is already in nature. Mathematics can be beautiful because it can capture the hidden patterns of nature, and nature itself can be beautiful.

To Cox, too, it came as no surprise to discover visual beauty in mathematics. "We say something is beautiful if it pleases the senses," she observes. "Now, how can something totally abstract as a mathematical idea be beautiful? Our mind simply has the same remarkable sensations when something is logically clear and finally fits together and you understand how it works—you say it is beautiful."

Cox remarks, "Art has been mathematicized as a result of computers, and those artists who really understand geometry and mathematics can use the computer to make some very unusual forms."

At the right, the Cox-Francis Etruscan Venus figure morphs from its "classic" form into three other topologically equivalent shapes.

MATHEMATICAL SYMPHONIES IN THE FOURTH DIMENSION

One of the earliest pioneers in computer graphics was Brown University mathematician Tom Banchoff. In the early 1970s, Banchoff worked with computer scientist Charles Strauss to develop a computer system that could draw geometric figures on a screen, when fed with the algebraic equations that described the figure symbolically. With Banchoff and Strauss's system, you no longer had to be a trained mathematician in order to see the beautiful shapes beyond the symbols; those shapes could be displayed on a computer screen for everyone to see. For the first time, mathematics was in a similar position to music, where a musical instrument can be used to turn the symbolic representation on a page of music into a melodious sound that everyone can hear.

For Banchoff himself, however, his new system was not a mathematical equivalent of a piano or a guitar. As an expert in geometry, he had no trouble "sight-reading" mathematical notation to "see" the geometric figure in his mind's eye. No, his interests lay elsewhere. For Banchoff, the new computer system was a transportation device in which he would set out to explore a new universe.

Ever since he was a child, Banchoff had been fascinated by the idea of the fourth dimension—a realm beyond our everyday experiences. The ordinary world in which we live has three dimensions—objects have height, width, and depth. Try as we may, we cannot picture a fourth dimension, perpendicular to the other three. And yet, physicists tell us that the universe we live in might actually have ten dimensions, seven of which are invisible to us.

Using mathematics and the computer, Banchoff set out to break free of our three-dimensional prison and explore—in mind if not in body—that unknown world of the fourth dimension.

To try to get a sense of dimension, Banchoff suggests the following exercise in mental visualization. Start by imagining a single point in space. That point has no dimension. Now move the point to a new position. The motion of the point traces out a line: a one-dimensional object. Next, move the line perpendicular to itself to trace out a square: a two-dimensional object. Translate the square in a direction perpendicular to all its edges and you trace out a cube: a three-dimensional object. Now do the same thing again. Translate the cube in a direction perpendicular to all its edges to trace out a hypercube, a four-dimensional "cube."

It's that last step that boggles the mind. The human mind has no trouble imagining the first three stages. But we can't visualize the fourth step. (In many diagrams of the hypercube, the fourth dimension is drawn as a diagonal.)

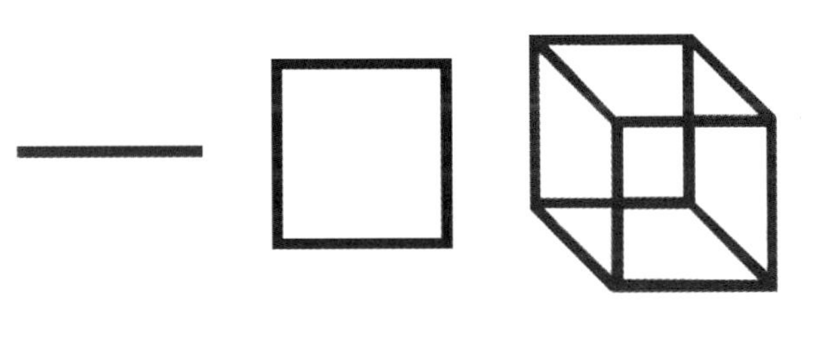
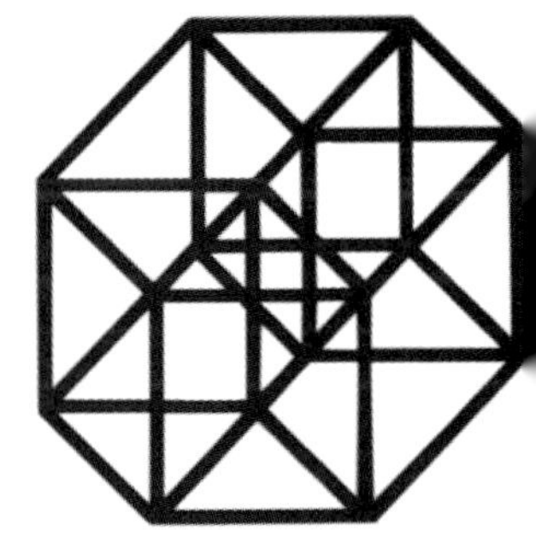

We are accustomed to the geometry of the line, the square, and the cube, but in the progression to the four-dimensional hypercube, the structure becomes mind-boggling.

Thought experiments like those described above had been around for many years, as had various line drawings similar to the images on this page. But Banchoff found them very unsatisfying. They didn't really help him visualize a hypercube. What he wanted to do was to walk around a hypercube, to see it from all angles. After all, Banchoff points out, that is exactly what we do when we want to really get a sense of a three-dimensional object. "We never really see a three-dimensional object all at once," Banchoff says. "If you're looking at a cube, a block, you only see half of it. As you walk round it, you see it from all angles and you piece it together."

Using computer graphics, Banchoff hoped to be able to "walk around" a hypercube. In 1978, he did just that. His childhood dream came true. For the first time ever, a human being got to see a genuine four-dimensional object—at least, "genuine" in the sense of mathematics. What Banchoff did was provide the computer with a complete and accurate description of a hypercube in the language of symbolic algebra. That part was easy. Trapped in three-dimensional space, our minds cannot fully grasp a four-dimensional object, but for abstract mathematics there is no such problem. You can write down precise mathematical descriptions of objects in four, five, six—any number of dimensions.

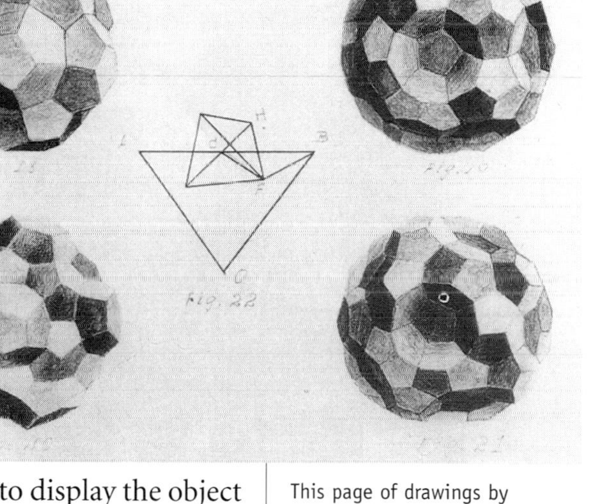

This page of drawings by American mathematician William Stringham, published in 1880, includes the first published drawing of a hypercube—second row down, in the middle. Although Stringham has often been credited with having discovered the hypercube form, in fact the Swiss mathematician Ludwig Schlaffli first worked out the hypercube's shape in a paper written in 1850.

Once Banchoff's computer contained a representation of a hypercube, the last step was to instruct the computer to display the object on the screen from all possible angles. Then Banchoff and his colleagues could have the experience of "walking around" the object. These days, the technology required for that second step is routine. Even the simplest of computer games will have a sufficiently sophisticated graphics display. But at the time Banchoff was trying to explore the hypercube, the graphics required was a leading-edge computer technique that he and Charles Strauss had to work out for themselves.

Over twenty years have passed since Banchoff caught his first glimpses of a hypercube. What are his recollections of that first moment? And have the intervening years enabled him to develop a better sense of what it looks like? Banchoff reflects: "When we saw the hypercube for the first time, we knew we were seeing something totally different. It moved in ways that were totally separate from anything we'd seen before. We were utterly taken by it. I realized I'd never completely understand it. The fourth dimension is beyond us, as much as the third dimension would be above people living on a two-dimensional plane."

Hypercubes are not limited to four dimensions, as seen above. Mathematicians have built on Thomas Banchoff's modeling of four-dimensional hypercubes to create ever more elaborate representations of higher-dimension cubes, as demonstrated here by a six-dimensional hypercube, at right, and an eight-dimensional hypercube, below.

But does the fourth dimension really exist? From a mathematical standpoint, the answer is absolutely yes. The hypercube has exactly the same kind of existence as any other geometric figure: a point, a line, a plane, a cube, or whatever. They all exist in the world of mathematics—in the invisible universe. As Banchoff says, "You can ask whether a square exists. There are no perfect squares, but we all have a conception of a square and we can recognize a representation of a square even though we know it's not perfect. But the real square is an abstraction, something in the mind of God, the Greek mathematicians would have said."

Other than pure curiosity, is there any reason to study the fourth dimension? Banchoff offers the following explanation: "One of the best advantages of studying higher dimensions is that it makes you much more sophisticated about your own. I see geometry in everything. When I walk down the street, I'm always fascinated by some form. I love seeing patterns; I love looking up and looking at windows and ledges and seeing the way different objects fit together, especially if you position yourself in just the right way."

For Banchoff, the same process of looking at things from different angles allows him to "see" a four-dimensional hypercube. "We learn to know a shape by seeing lots of different views, and viewing them in our minds in some sort of network of associations," he says. "By that same sort of process of exploration, you can get to know a four-dimensional shape. I see that again and again with my students over the course of a semester."

"Geometry is about the kinds of structures and relationships that we can try to visualize and see. We can see some of the same elements in theology and in religion. It's very difficult to imagine some transcendental being without using prepositions like 'above' that make us think in spatial terms."

TOM BANCHOFF
geometer

A DIFFERENT VIEW

Though Tom Banchoff was one of the first people to get a truly good view into the fourth dimension, he was by no means the first person to try. Artists and writers had been fascinated by the idea since the end of the nineteenth century. Much of the interest was spurred by developments in mathematics and science that had spilled over into popular culture.

During the nineteenth century, mathematics became much more abstract than ever before, as mathematicians began to explore new geometries—geometries in which parallel lines meet, geometries of spaces with four or more dimensions, even geometries of spaces with infinitely many dimensions. At the same time, with the discovery of X rays, scientists gave people the ability to see the bones inside their own bodies.

According to art historian Linda Dalrymple Henderson, the new developments led to a change in worldview every bit as significant as that caused by the discovery of perspective in the Renaissance era.

At right, the first published X-ray photograph, which was reproduced in Wilhelm Röntgen's paper announcing his accidental discovery of X rays, in 1896.

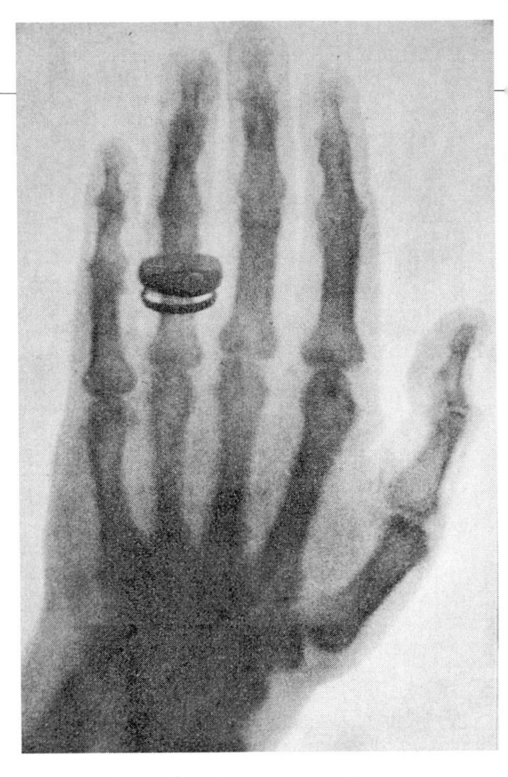

In the cubist painting below, "Man with a Guitar," painted in the spring of 1913, Pablo Picasso makes dramatic use of new techniques for representing objects in space. By violating the established rules of perspective and inventing a new language of representation, the cubists sought to depict a more multidimensional view of the world than that afforded by normal human vision. Photograph ©1997 The Museum of Modern Art, New York. André Meyer Bequest.

The appearance of X-ray photographs of the interior of the hand, Henderson tells us, "caused an incredible reverberation in culture. The popular response is remarkable: hundreds of articles and books, poems and songs. This is really a remarkable moment that we tend to have forgotten."

In painting, the mathematician's new interest in the geometry and structure of space led to the cubist movement, where artists such as Pablo Picasso sought ways to explore and represent dimensionality. "It is a marvelous utopian moment," says Henderson, "a kind of feeling that this is a new world, a new century. We must have new languages in which to embody it: verbal languages, visual languages, mathematical languages."

Tony Robbin is a present-day artist who continues the tradition of trying to find ways to look beyond our three-dimensional universe. Like Tom Banchoff, Robbin has always been particularly fascinated by the idea of the fourth dimension. Whereas Banchoff explores four-dimensional space as a mathematician, Robbin tries to view it through the eyes of an artist.

Robbin explains his fascination with exploring four-dimensional space in painting this way: "I think the main vehicle that artists use to communicate emotion is space. You can say the history of art is the history of different spaces, and these spaces are an embodiment of the mathematics of the time, of the geometry of the time. What four-dimensional geometry can do is give you a model of the complexity of the world we live in."

"I think the main vehicle that artists use to communicate emotion is space. . . . What four-dimensional geometry can do is give you a model of the complexity of the world we live in."
TONY ROBBIN
artist

Though Robbin has always been fascinated by dimension, it was not until he met Banchoff, and saw the hypercube animation, that he felt he could really "see" the fourth dimension. "For days I dreamed only of these images rotating and turning inside out and behaving in these very mysterious and yet recognizable ways," he says.

For Robbin, as for Banchoff, the challenge was to find ways to visualize the fourth dimension. In Banchoff's case, mathematics was the medium. For Robbin the challenge was an artistic one: "The problem for the artist who wants to make art about space is to make space itself present, to make space itself visible.

"For thirty thousand years," says Robbin, "human beings have made patterns. It's a fundamental way we have to organize and structure our experience. So I think that making patterns becomes the ideal way to define space."

Tony Robbin sketching the outlines of a painting in which he makes use of four-dimensional geometry.

Robbin creates his four-dimensional artwork by representing the shadows of a four-dimensional world on a two-dimensional canvas, enhancing the effect by building out from the canvas with nails and wire—bringing the canvas itself out into the three-dimensional world we live in. The key to giving the viewer a "genuine" view of the fourth dimension is to combine the patterns on the canvas and the pattern of the nails and wire so that, by moving around in front of the canvas and viewing it from a changing perspective, the viewer

Untitled #20, by Tony Robbin, 1978.

begins to experience what it would be like to be in a four-dimensional world. For this to work, Robbin has to get the geometry of a four-dimensional world right. And that means he has to begin with the mathematics—with the geometry.

As for Banchoff, the computer provides the key. For, unlike people, computers operate in a symbolic world of unlimited dimensions. These days, you can go to the computer store and buy graphics programs off the shelf, but when Robbin was starting his artistic exploration of the fourth dimension, there were no such programs. He had to go back to school to learn enough mathematics and computer programming to write his own software. "It was a tremendous task," he says, "but it was worthwhile. I really learned a lot of mathematics by doing this."

For Robbin, then, the computer is a crucial tool for the artist who wants to examine reality. "I think that the computer is the biggest thing since the lens," he says. "The lens allowed us to see things that were far away up close, and allowed us to make recorded images of the things we saw. The computer allows us to see things that we know are there but we can't see for ourselves. It creates a visual world of four-dimensional geometry, of hyperbolic geometry, of quasi-crystal geometry, of fractal geometry. We know that these structures exist. The computer allows us to see them."

How real is the experience that an artist like Robbin can give us? Just because he starts with the mathematics, does that ensure that the artwork he produces represents the "real" four-dimensional world that the geometry defines? After all, filmmaker Doug Trumbull also uses computer graphics to produce images on a screen, but unlike Banchoff, Trumbull creates worlds that are purely imaginary. For Trumbull, the mathematics is just a means to an end; it provides him with the tools to create his imaginary worlds.

Banchoff, on the other hand, does not create worlds; he explores a world that already exists—that exists within the realm of mathematics. Like other mathematicians, Banchoff does not worry about whether there really is a fourth dimension. Using mathematics, he can describe what the fourth dimension would look like if it exists. This is one of the most fascinating aspects of mathematics: it can show you what something would be like, even if it doesn't really exist.

Robbin is somewhere between Trumbull and Banchoff. He sets out to explore the mathematically real world of the fourth dimension. But he wants to use his artist's abilities in order to interpret and represent that world. So how can we be sure that what Robbin shows us is truly representative? Robbin agrees there is a problem here, but it is not, he insists, a problem to do with the fourth dimension. The problem is to do with seeing in the first place. "I think we have a misunderstanding about seeing," he says. "We think we see the world out there with our eyes and that this process is objective and automatic. In fact, we see with our minds. We go through life doing geometry. We go through life applying the geometry that we know on our visual world. If something doesn't fit that geometry, we don't see it."

BEYOND PERSPECTIVE

Today, mathematicians and artists are starting to venture beyond the realms of Banchoff's computer screen and Robbin's canvas. Using virtual-reality technology, people can not only view worlds of four or more dimensions, they can experience them, flying through them and viewing them from all angles.

Marcos Novak viewing some of his higher-dimensional creations with virtual-reality goggles.

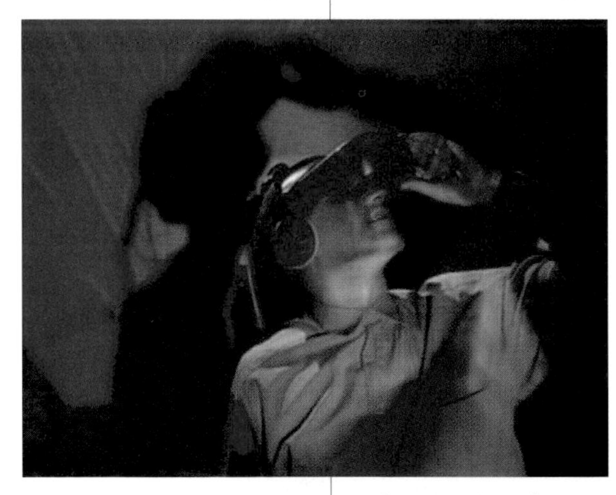

Art instructor Sam Edgerton describes the new developments this way: "The Renaissance perspective artist was constantly provoked by his patrons: 'Give me something more dramatic. I want to see something even more perspectival.' Well, we've already transcended that. Today we are into virtual space and so on, and I don't know how far we can go with this."

Marcos Novak is a pioneer in the area Edgerton is referring to. Many would say he is a "super-Renaissance artist," but Novak himself prefers the term *architect*—someone who designs spaces for people to live in, to work in, or to play in. Unlike most architects, however, whose ideas are eventually turned into physical, three-dimensional structures, Novak concentrates on the

This complex image is an architectural composition by Marcos Novak. To create the image, he has taken a three-dimensional structure and first transformed it into the fourth dimension and then projected it back into the third dimension, creating a perpetually changing "liquid" form to be viewed in virtual reality. Novak has coined the term *transarchitecture* to describe such designs, which break the bounds of architecture as we know it.

creation of virtual environments, often of many dimensions. Those worlds will never be turned into bricks and stone, but they can be experienced. Always under development and subject to revision, they are, to use Novak's term, "worlds in progress."

Put on video goggles and enter Novak's virtual universe, and you will find yourself in a world unlike anything you have experienced before. You can find yourself traveling through the three-dimensional shadow of a four-dimensional hypercube, free to turn around and view it from any angle you choose. Go through a doorway and you are in another world, a world filled with flying creatures, only these creatures are unlike any birds you will find in the everyday world. These creatures are colorful geometric shapes that flap their geometric wings—Novak quips they might be called "bats of paradise."

"People often ask me why I'm building these strange things," Novak says. "It seems to me that if you try to imitate reality, it's doomed to fail, because as a replica of reality, it's not very good. I'm trying to

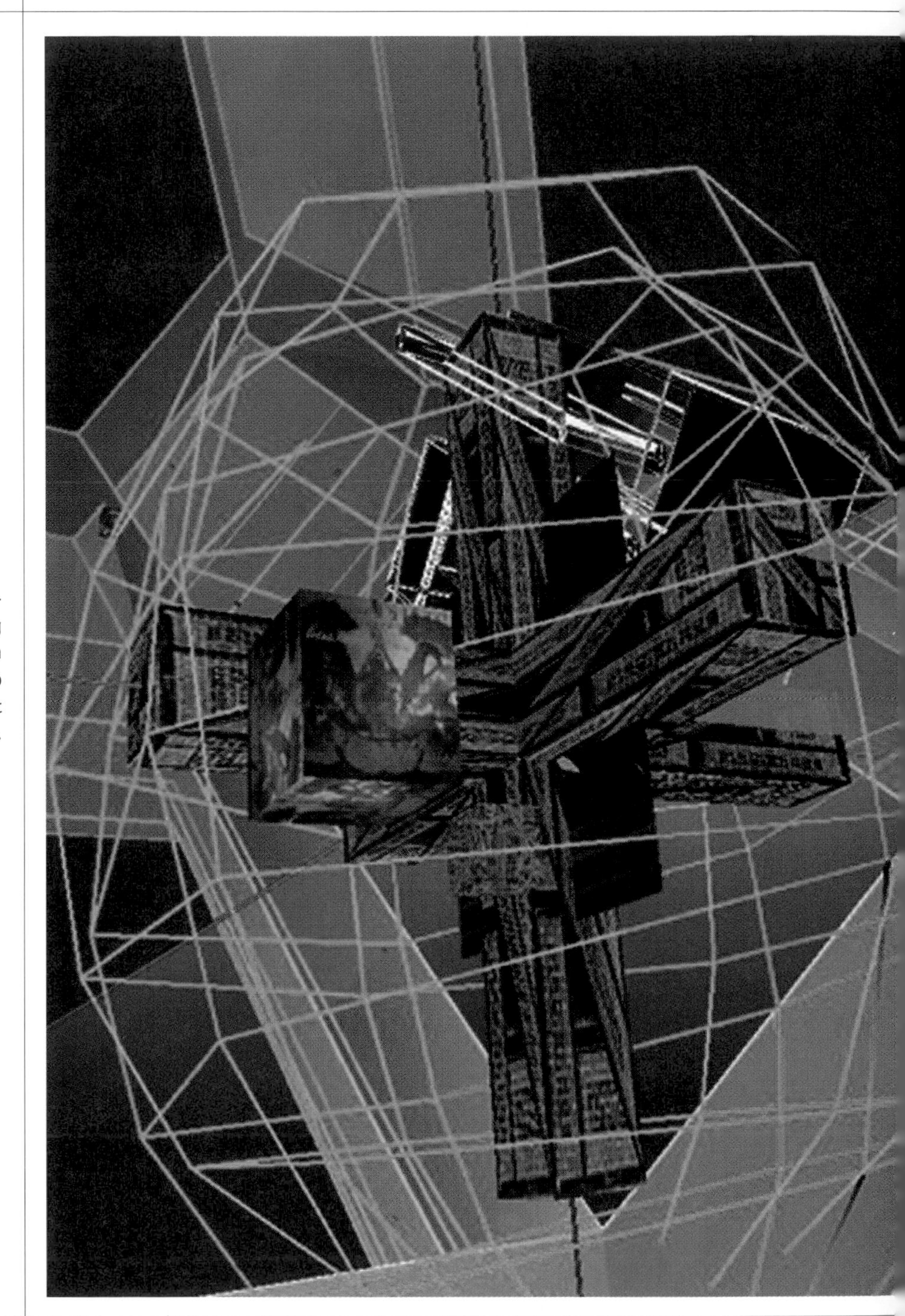

This computer drawing represents a four-dimensional building designed by Marcos Novak, in which there is no traditional top and bottom or left and right sides to the structure.

find out what new things are possible. I set out to create a vast unexplored territory we can venture into and, I'm sure, come back from with incredible insight and beauty."

Novak starts out by acknowledging that when we experience the outside world—when we see, touch, hear, smell, or feel—we do so with our inner consciousness. The world remains outside but our experience of it is internal. We each have our own experience, shaped by who we are. If we want, we can use our imagination to modify that experience—to imagine things as slightly different from the way they appear to us. Novelists, in particular science fiction writers, use words to convey those imagined worlds to others. Novak can do much more than that. By translating his inner experiences—his imagined worlds—into the language of mathematics, he can create that inner world in a computer so that others can experience it. This is, Novak says, a completely new way to communicate our inner selves.

As Novak puts it, "You can argue that our consciousness itself comes from our ability to create a pocket of virtual space inside our minds that's a mirror of the actual space outside. What I try to do is take this mirror of the world which is in the seat of my consciousness and put it back outside my body, in the public realm. I can give it to you to share. And you can do that as well. What mathematics does is give you a concise and accurate way of sharing our inner world. I think it's absolutely astonishing and beautiful that we can do that."

"I think we can never know enough mathematics," Novak adds. "Mathematics is not about numbers. That's what arithmetic is about. Mathematics is about structures—possible structures. And virtual reality is about visiting possible worlds."

"Most people associate architecture with built materials: wood, brick, mortar, and stone. The things I do involve building architecture out of information."

MARCOS NOVAK
virtual-reality architect

Chapter 3

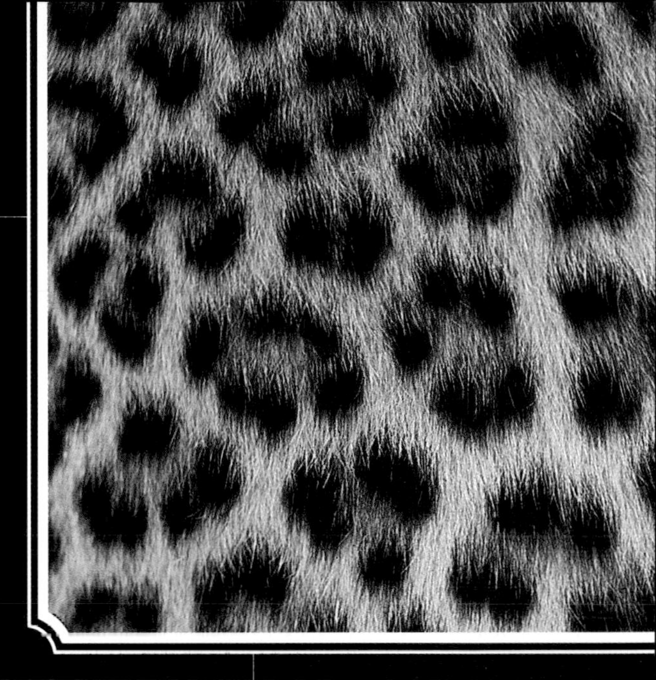

PATTERNS OF NATURE

Why are animals the sizes they are? Could people grow to be twelve feet tall? Could centipedes be large enough for us to ride on them as we ride on horses? (Think of the increased efficiency that would result, with several people sharing a ride on a single centipede.) How do animal coats develop their distinctive patterns?

How do flowers and plants "know" how to grow the way they do? How do species of animals and plants really evolve? These are all questions about the world we live in. They begin with what we can see, but they ask about the hidden things we can't see—the why and the how. To provide answers, we have to find a way to make the invisible visible.

In the previous chapter, we saw how mathematics can make visible to others the invisible products of our imagination. Now we'll see how mathematics can help us make visible to ourselves the hidden secrets of nature.

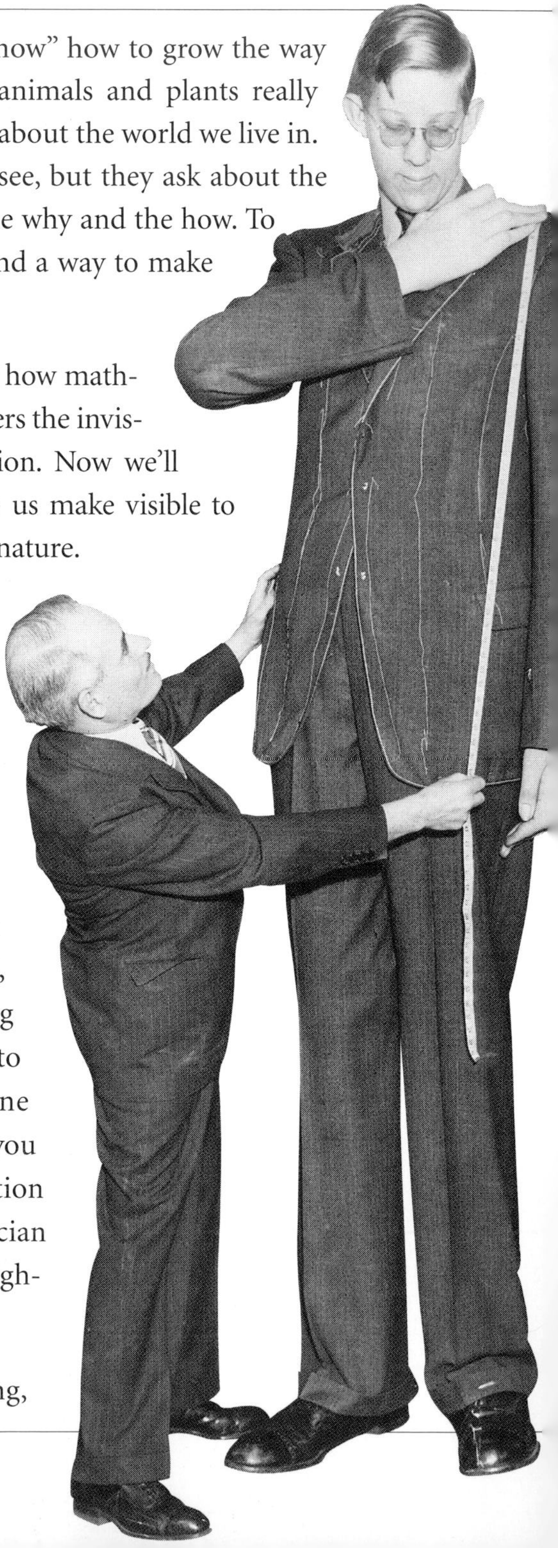

The tallest person on record, Robert Wadlow, stood eight feet, five inches tall. He is shown here being sized for a new suit by tailor Sol Winkelman in 1936.

For example, without mathematics, there is no way you can understand what keeps a jumbo jet in the air. As we all know, large metal objects don't stay above the ground without something to support them. But when you look at a jet aircraft flying overhead, you can't see anything holding it up. It takes mathematics to "see" what keeps an airplane aloft. In this case, what lets you "see" the invisible is an equation discovered by the mathematician Daniel Bernoulli early in the eighteenth century.

Staying on the subject of flying,

what is it that causes objects other than aircraft to fall to the ground when we release them? "Gravity," you may answer. But that's just giving it a name. It doesn't help us to understand it. It's still invisible. We might as well call it magic. To understand it, you have to "see" it. That's exactly what Newton did with his equations of motion in the seventeenth century. Newton's mathematics enabled us to "see" the invisible forces that keep the earth rotating around the sun and cause an apple to fall from the tree onto the ground.

Here's another example: How do you "see" what makes the pictures and sound of a football game miraculously appear on a television screen in another part of the country? One answer is that the pictures and sound are transmitted by radio waves—a special case of what we call electromagnetic radiation. But as with gravity, that just gives the phenomenon a name; it doesn't help us to "see" it. In order to "see" radio waves, you have to use mathematics. Maxwell's equations, discovered in the last century, make visible to us the otherwise invisible radio waves.

These examples focus on revealing the hidden laws of the physical world. They all use mathematics developed in former days. In contrast, the examples described in the remainder of this chapter are all part of a quest to discover the patterns of life, and they are all new. This work is so recent that most of the people who pioneered these new applications of mathematics are alive today. As you read on, you will meet some of these new pioneers, and learn about the way they are making visible the hitherto invisible patterns of the living world around us.

ALL CREATURES GREAT AND SMALL

Biologist Mike Labarbara enjoys old science fiction movies. But like many scientists, occasionally he finds that the action on the screen

King Kong atop the Empire State Building in the classic 1933 movie.

sets his scientist's brain into motion, and instead of following the plot he starts to wonder about the science involved. He sees this as almost inevitable. "One important aspect of being a really good scientist is keeping curiosity alive and uncensored into your adult life," he says. But he acknowledges that it can make it hard for a scientist to watch certain movies, especially when they involve scientific issues the scientist is expert in. In Labarbara's case, the problems arise with movies that involve giant creatures—soldiers fighting giant locusts, cities being overrun by giant ants, the heroine being swept away by the giant gorilla King Kong, and so forth. For Labarbara is an expert in animal size and strength.

Though animals come in many shapes and sizes, there are definite limits on the possible size of an animal of a particular shape. King Kong simply could not exist, for instance. As Labarbara has calculated, if you were to take a gorilla and blow it up to the size of King Kong, its weight would increase more than 14,000 times but the size of its bones would increase by only a few hundred times. Kong's bones would simply not be able to support his body. He would collapse under his own weight!

And the same is true for all those giant locusts, giant ants, and the like. Imagining giants—giant people, giant animals, or giant insects—might provide the basis for an entertaining story, but the rules of science say that giants could not happen. You can't have a giant anything. If you want to change size, you have to change the overall design.

The reason is quite simple. Suppose you double the height (or length) of any creature, say, a gorilla. Then the weight will increase 8 times (i.e., 2 cubed), but the cross section of the bones will increase only fourfold (2 squared). Or, if you increase the height of the gorilla 10 times, the weight will increase 1,000 times (10 cubed), but the cross-sectional area of the bones will increase only 100 times (10 squared). In general, when you increase the height by a certain factor, the weight will increase by the cube of that factor but the cross section of the bones will increase only by the square of that factor.

Nine squares are required to produce a square three times the size of a single square. Twenty-seven cubes are required to produce a cube three times larger than the single cube.

This simple relationship between the increase in length, area, and volume (and hence weight, since weight depends on volume—on how much material there is) is most easily imagined by thinking of

Biomechanist Mike Labarbara atop a chair fit for a man more than twice his size.

a sugar cube. Suppose you wanted to make a "giant" sugar cube, three times the size of the original cube, by sticking together lots of ordinary sugar cubes. To make a sugar cube a mere three times the size of the original one, you need to stick together a total of 27 cubes! The reason is that the new cube does not simply have to be three times wider than the original one; it also has to be three times deeper and three times taller. That means you need $3 \times 3 \times 3 = 27$ cubes altogether.

A cross section of the giant sugar cube, on the other hand, has only nine ($= 3 \times 3$) cubes in it. So cross section increases faster than height (9 times faster in the case of the sugar cube), but

volume (and hence weight) goes up much more rapidly (27 times for the sugar cube).

This enormous differential between the increase of volume/weight and the increase in cross section would not prevent the growth of giants if the bones of an animal were made of a sufficiently strong material. But as scientists have recently discovered, they are not. In fact, every time we move, we and all other creatures of the earth come very close to breaking our bones! As Labarbara explains: "In mammals, the bones and muscles are put together in such a way as to always keep the maximum stresses that are exerted on the bones about a quarter to a third of the breaking stress of the bone. This applies across the board, from mammals as small as chipmunks to as large as elephants."

Labarbara points out that no engineer would design a bridge or a building with such a low safety threshold. "Engineering structures are never built to exert stresses so close to failure as we see used routinely in mammals," he says. Because of this low safety threshold, when humans and other animals want to go faster, they change the way they move, in order to reduce the stress on the bones. With the aid of mathematics, Labarbara has been analyzing the way animals move.

"Mathematics has to become a bigger part of biology. We're just beginning to understand how animals and plants are designed."
MIKE LABARBARA
biologist

For example, look at the way a horse moves, says Labarbara. When it starts to walk, at first there is little stress on the bones. The faster it walks, the greater the stress. Then, just when the stress level gets up to around 30 percent of breaking point, the horse changes the way it moves, from a walk to a trot. Immediately, the stress level falls off; the bones are no longer in danger. As the trotting speed increases, the stress level starts to climb again, until it once again approaches the critical 30 percent level, and then the horse breaks into a canter. At once the stress level falls, relieving the pressure on the bones. Finally, when the cantering speed increases enough to bring the stress level once again to the 30 percent level, the horse changes its gait to a gallop.

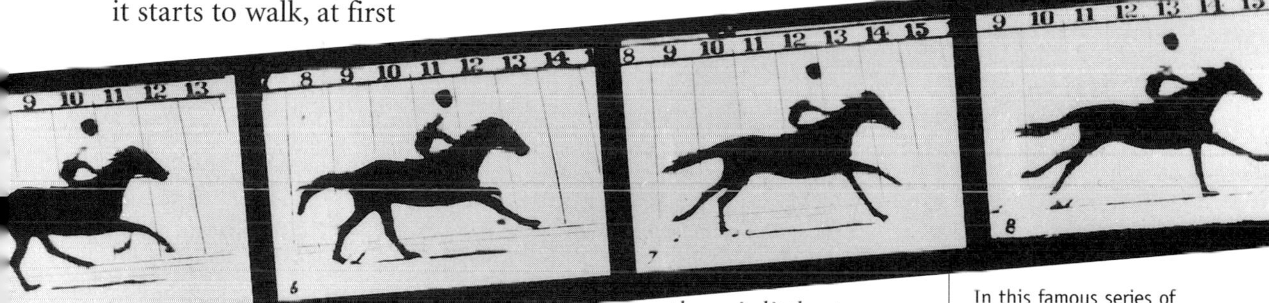

In this famous series of photographs by Eadweard Muybridge, made in 1878, titled "The Horse in Motion," Muybridge proved that a horse lifts all four hooves off the ground in mid-gallop. This is most clearly seen in the third frame of the series.

You see the same general pattern with humans: walking, jogging, running, and sprinting, each involving a very different kind of leg, foot, and body motion.

Labarbara uses mathematical models to help analyze and explain the way animals such as horses move. A mathematical model is just a collection of graphs and equations that describe some activity or phenomenon in the language of mathematics. By studying slow-motion film of, say, a horse in motion, and comparing the film with laboratory data on the bones and the body structure of the horse, Labarbara can develop a mathematical model of horse movement. By comparing the models of motion for different animals, he can make general conclusions about animal movement—he can develop a general mathematical model.

Had the *Tyrannosaurus rex* walked upright as once thought, the weight of its massive head would have crushed the vertebrae of its spine.

Using a mathematical model, Labarbara has investigated the way extinct creatures such as dinosaurs must have moved. In this case, he has to use the mathematical model "backward." He doesn't have any film of a dinosaur in motion. Rather, he wants to find out what it must have looked like when it moved. To do this, he starts with the laboratory data on the skeleton and the bone structure that scientists have produced by examining the bones found at archaeological sites. He then uses his mathematical model to deduce the way the animal must have moved, assuming the "30 percent rule."

Labarbara has used this technique to study the tyrannosaurs. "Rather than being birdlike in the way they moved," he observes, "they were much more like mammals, with their legs held under the body and straight during locomotion."

For Labarbara, mathematics has provided a way to look back in time to the days when dinosaurs roamed the earth. "Mathematics is changing the way we understand these animals as having held their bodies," he comments, "and is thus changing the way we think about how they hunted and lived. The more you know about mathematics, the deeper you'll see, and the more you'll be able to extract the underlying structure and beauty."

This painting of Jurassic-era dinosaurs depicts the latest thinking about the way in which many dinosaurs must have carried their bodies as they walked and ran.

HOW THE LEOPARD GETS ITS SPOTS

When the young daughter of University of Oxford mathematician James Murray asked him how a leopard gets its spots, he did not know the answer. Nor did any of his scientist colleagues at Oxford. That was over thirty years ago. Today, Murray, now at the University of Washington in Seattle, has a possible explanation.

The problem was not one of chemistry. Scientists have known for many years that skin coloration is caused by a substance called melanin produced by cells just beneath the surface of the skin. The question was, What makes the cells produce melanin in the form of a characteristic pattern—spots in some animals, stripes in others, and so forth?

Right, a close-up of the skin of a cougar, and opposite, the skin of a leopard. James Murray wondered what could create two patterns that are so similar and yet so distinctively different.

Murray found a mathematical answer. Just as Labarbara constructed a mathematical model of animal motion, so Murray developed a mathematical model of the production of melanin in the skin of an animal. Whereas Labarbara's model was based on physics, Murray's model depended on chemistry—on the laws that govern diffusion and chemical reactions.

Murray began by assuming that certain chemicals stimulate the cells to produce melanin. According to this assumption, the visible coat pattern is simply a reflection of an invisible chemical pattern in the skin: high concentrations of the chemicals give rise to a melanin coloration; low concentrations leave the skin largely uncolored. The question then was, What causes the melanin-inducing chemicals to cluster into a regular pattern formation so that when those chemicals "switch on" the melanin to turn color, the result is a visible pattern in the skin? This was the specific question Murray addressed.

He answered it by looking at so-called reaction-diffusion systems. A reaction-diffusion system is where two or more chemicals in the same solution (or in the same skin!) react and diffuse throughout the solution, fighting with each other for control of the territory. Though first proposed by mathematicians as a theoretical idea in the 1950s, reaction-diffusion systems were only later observed by chemists in the laboratory. Even today, they are still studied more by mathematicians, in a theoretical way, than by chemists in the laboratory.

To obtain his model, Murray assumed that two chemicals are produced in the skin, one of which stimulates the production of melanin, the other of which inhibits this effect. He further assumed that production of the stimulating chemical triggers production of the inhibitor. Such a system could in principle lead to the formation of "islands" of the stimulating chemicals surrounded by "fences" of inhibitors, causing the formation of melanin "spots." Here's how.

"It is really a very exciting time to be a mathematician who works in biology. I find it quite difficult not to look at a fern or the bark of a tree and wonder how it was formed. Why is it like that? There are so many questions we would like to know the answers to."

JAMES MURRAY
mathematician

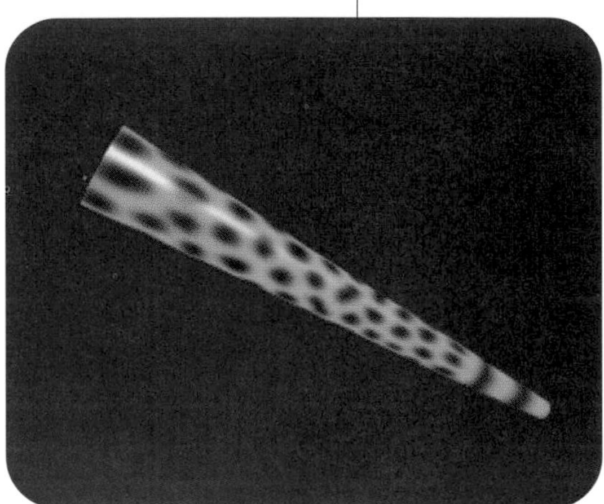

Murray found that by changing the equations in his computer models of animals' tails he could transform a spotted tail into a striped one.

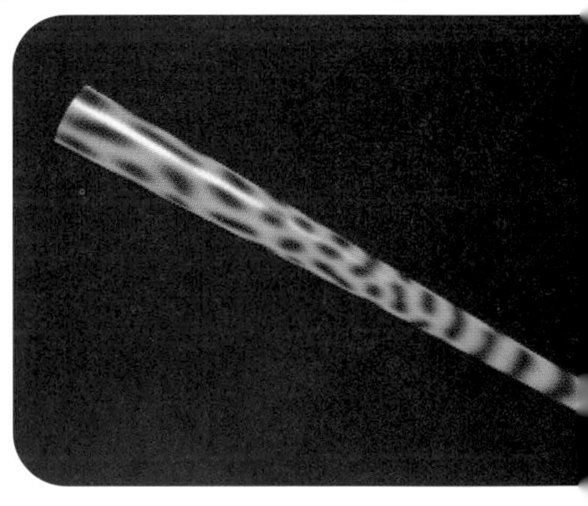

Suppose that the inhibiting chemical diffuses faster than the stimulator. If a concentration of the stimulating chemical is formed, triggering the production of the inhibitor, then the faster-moving inhibitor will be able to encircle the more slowly diffusing stimulator, preventing further expansion. The result is a "spot" of the stimulator, kept in check by an encircling ring of the inhibitor.

Murray likens this process to the following scenario. Imagine, he says, a very dry forest. Because of the danger of fire, fire crews are stationed throughout the forest with helicopters and fire-fighting equipment. When a fire breaks out (the stimulator), the fire fighters (the inhibitors) spring into action. Traveling in their helicopters, they can move much more quickly than the fire. (The inhibitor diffuses faster than the stimulator.) However, because of the intensity of the fire (the high concentration of the stimulator), the fire fighters cannot contain the fire at its core. So, using their greater speed, they outrun the front of the fire and spray fire-resistant chemicals onto the trees. When the fire reaches the sprayed trees, its progress is stopped. Seen from the air, the result will be a blackened spot where the fire burned, surrounded by the green ring of the sprayed trees and beyond it the green of the remainder of the forest.

Now imagine what happens if a number of fires break out all over the forest. Seen from the air, the resulting landscape will show a pattern of patches of blackened, burned trees interspersed with the green of the unburned trees. If the fires break out sufficiently far apart from one another, the resulting aerial pattern could be one of black spots in a sea of green. On the other hand, if nearby fires are able to merge before being contained, different patterns could result. The exact pattern will depend on various factors, in particular the number and relative positions of the initial fires and the relative speeds of the fire and the fire fighters (the reaction-diffusion rates).

The case of interest to Murray was what the resulting pattern would be if the initial fire pattern were random. In that case, how would the different "reaction-diffusion rates" affect the final pattern? More specifically, were there rates that, starting from a random pattern of fire sources, would lead to recognizable patterns, such as regular spots or stripes?

This is where the mathematics came in. Chemists and mathematicians can write down equations to describe the way chemicals react and disperse. (They are a special kind of equation called partial differential equations, which involve techniques from calculus.)

As a first step, Murray picked the easiest case he could think of that might lead to a regular coat pattern: he assumed there were just two chemicals, and that they reacted and diffused at different rates. Having made his assumptions and written down his equations, the rest was entirely mathematics. Putting his model into a computer, Murray was able to turn his equations into pictures on the screen, showing the way the react-

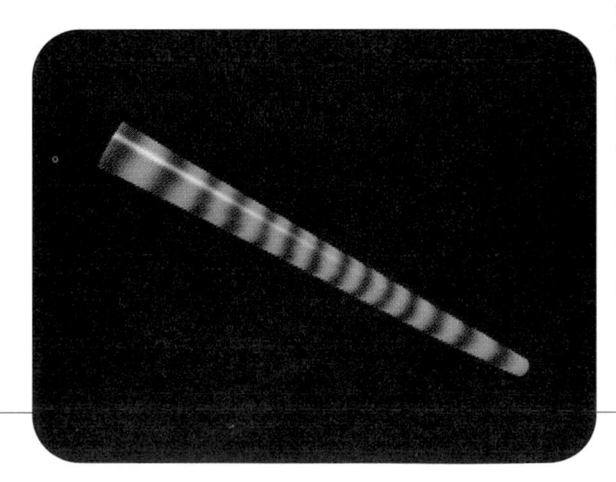

"Mathematics in the biomedical sciences is like a sophisticated laboratory tool which can be used to suggest other questions or to generate hypotheses."
JAMES MURRAY
mathematician

ing chemicals dispersed. To his surprise, even with just two chemicals, his mathematical model produced dispersal patterns that looked remarkably like the patterns seen on the skins of animals.

That was Murray's first success, and it confirmed his initial suspicion that the answer to his daughter's question might be found by looking at reaction-diffusion systems. But the story does not end there. What Murray did next led him to make a remarkable and quite unexpected discovery.

By changing the values of certain numbers in his equations that correspond to the area and shape of the skin region, Murray found that his model produced different kinds of patterns. For very small regions, there was no pattern at all. For larger regions he got stripes, small spots, large spots, et cetera, reminiscent of the stripes of the leopard's tail, the stripes of the zebra, the smaller spots of the cheetah, the larger spots of the giraffe, and so forth. For very large regions, he got no pattern at all.

He also found that if he set up his model to correspond to an area large enough to sustain spots, and if he then changed the model to make the skin area very long and thin, but with the same overall area, then the spots always turned into stripes. In other words, his equations predicted that if the skin area of the animal was sufficiently thin, it couldn't have spots, only stripes. In the case of real animals, this would imply that you could have a spotted animal with a striped tail, but never a striped animal with a spotted tail. Which is exactly what we find in nature—the leopard and the cheetah are good examples of spotted animals with striped tails. Amazingly, Murray's simple equations were predicting exactly the kinds of skin pattern combinations that actually arise.

Thus, Murray's mathematical model shows that one very simple mechanism can produce all the familiar markings we see on many different animals. To get different kinds of patterns, all Murray had to do was change the value of certain numbers in his equations—numbers that corresponded to the size and shape of the region on which the pattern was being produced.

Assuming his model reflected what really occurs, Murray's next question was a biological one: What would cause different kinds of patterns to form on different species of animals? For example, why does the leopard produce spots but the tiger stripes? Since leopards and tigers have roughly the same size and shape, the answer cannot lie in the final, adult form. It must, Murray reasoned, have to do with embryonic development.

More precisely, if the kind of pattern depends on the size and shape

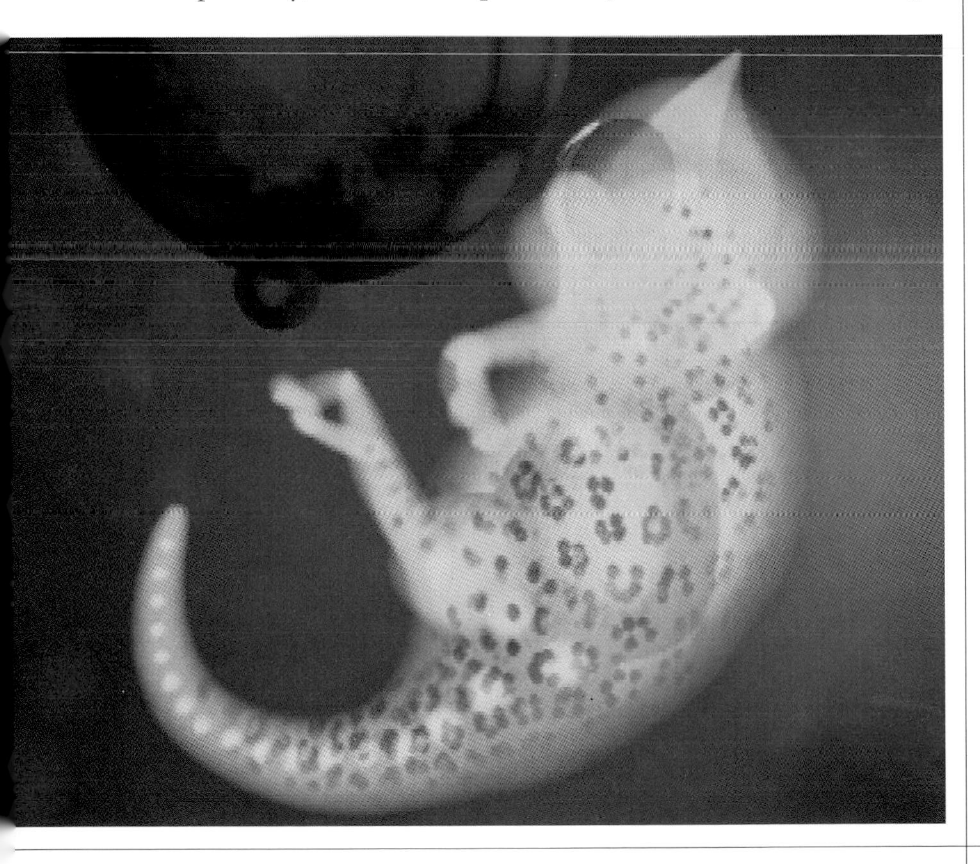

A computer animation of a leopard embryo developing its spots.

of the skin region, as Murray's model suggests, then the difference in patterns must be a consequence of when the chemical reactions take place during the animal's early maturation process. That is, the pattern we see on the adult animal will depend on the size and shape of the embryo when the process occurred.

According to the equations, the larger the embryo when the process occurs, the more complex the patterns that result, up to a point: with increasing size of embryo, the pattern changes from no pattern to stripes to spots and then back to no pattern again.

For example, in the case of mice, if the chemical reactions occur when the embryo is very small, then, according to Murray's mathematics, no pattern is possible. For a zebra, there is a four-week period early in the year-long gestation during which the embryo is long and pencil-like. Murray's mathematics predicts that, if the reactions take place during this period, the resulting pattern will be stripes.

More predictions: The genet has a long, thin tail throughout gestation, and as a result its tail is all striped. The leopard's tail is short and stumpy for much of its embryonic development, and as a result the spots can form along part of the tail, before it becomes so thin that you get only stripes.

It's all very intriguing. But is it right? Does Murray's model correspond to what really goes on? So far, we don't know. What we do know is that his mathematical equations for reaction-diffusion do lead to the kinds of skin patterns we see in nature. Moreover, no one has yet produced a better alternative. For Murray, who believes that nature usually favors efficiency, the extreme simplicity of his model is suggestive: "The idea that a single mechanism could produce all the coat patterns observed in the mammalian kingdom seems to me a very efficient process."

But, regardless of whether his mathematical model turns out to correspond to what actually happens in nature, Murray sees it as completely natural to have tried to use mathematics to explain how the leopard gets its spots. He sees the use of mathematics in biology as "the fastest-growing, most exciting area of mathematics that I can possibly conceive."

"Mathematics is not the mystery some people think it is," he says. "It's just a way of speaking.

"It's very imaginative."

FIGHTING VIRUSES WITH KNOTS

James Murray is just one of a number of scientists who have found ways to apply mathematics to biology. Another recent application is in the fight against viruses.

When mathematicians started to develop a theory of knots in the middle of the nineteenth century, fifty years were still to elapse before scientists would identify viruses as the cause of many of the illnesses that afflict us. Yet, as so often happens with mathematical theories developed initially out of pure curiosity, the mathematical theory of knots has turned out to provide biologists with a powerful tool in their quest to understand—and thereby, it is hoped, to conquer—viruses.

In many ways, knot theory provides an excellent example of mathematics as the study of abstract patterns. For one thing, knots clearly come in different patterns—there are different kinds of knots: overhand knot, figure-eight knot, et cetera. And yet, what distinguishes these knots from each other has nothing to do with the material the knot is made of. You can have the same knot, say, an overhand knot, made out of thread, string, electric cable, garden

hose, a necklace, a sock, a lock of hair, and so forth. The characteristic feature of a particular knot—what distinguishes it from other knots—is the pattern of the knot, the way it winds around itself.

To some extent, you can see the pattern of a knot when you look at one. On the other hand, what you see is just one particular "presentation" of the knot, one way of laying it out. Imagine taking an overhand knot tied in a piece of string and then laying it out in different ways: loose, tight, one big loop and one tight loop, arranged so that the string lies straight with sharp corners rather than in the usual smooth curves, et cetera. Everyone would agree that it is the same knot in each case—an overhand knot—tied in the same piece of string. It looks different in each case, but it is the same "knot." This is because the knot pattern itself is more abstract than what you actually see.

How do you go about studying knot patterns mathematically? First, you have to agree on some standard way to represent those abstract patterns. One way is by using simple line diagrams, much as in geometry. In the case of knots, the shape of the lines will not matter—they can be smooth curves or straight lines with corners,

An array of fascinating knots, representing just a small selection of the possible ways that knots can be "tied."

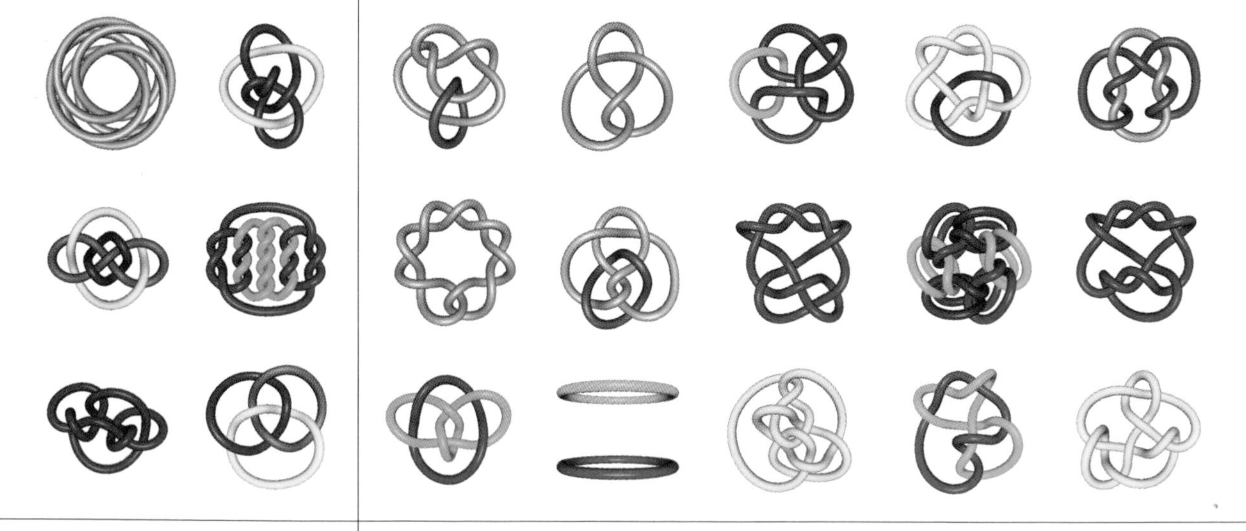

or anything in between. What counts is how the lines pass over and under each other. You can show that in a diagram by putting a small break in the line that passes underneath another one.

There is one technical stipulation that mathematicians have to make before they can begin the analysis of such diagrams: They require that when a knot has been tied, say in string, the two free ends of the string are fused together so that the string forms a closed loop. This prevents the knot from becoming untied. Once the ends have been fused together, the only way to untie a knot involves cutting the string. This requirement allows mathematicians to distinguish, in a formal way, between two knots that just *look* different and two knots that actually are different. With fused-end knots, a mathematician can say that two knots are really different if one cannot be manipulated into the other without cutting the string. If the ends were free, any knot could be manipulated into any other knot by first untying the initial knot and then retying it in the form of the other knot! Everyone would agree that such a maneuver would be "unfair" in knot theory. Joining the ends prohibits it.

Above, a trefoil knot in the "open" form and in the "closed" form, and below, a figure-eight knot in the open and closed forms.

By examining knot diagrams, mathematicians have been able to analyze the different kinds of knots that can be formed. This approach is helpful up to a point, but there are two problems with it. First, knot diagrams are really useful only for relatively small knots, for which a diagram can be drawn. They don't allow the mathematician to talk about "all possible knots." The second problem is that, except for the very simplest of knots that have fewer than five or six points where the line crosses over itself, it is hard to be sure that two knots whose diagrams look different really are different. When it comes to

"It's remarkable to discover how many connections mathematics makes, how many lightbulbs it suddenly turns on."
SYLVIA SPENGLER
biologist

knots, appearances can be very deceptive.

For example, anyone who has tried to unravel a tangled string of beads will know that it can be a highly frustrating task. And yet, provided the string has been put away with the clasp secure, it can't really be knotted at all. It went into the drawer as a simple, unknotted, closed loop, and so it must have stayed—unless someone came along and undid the clasp, knotted the string, and then refastened the clasp. The tangles you see do not actually create a knot in the string. It just looks like a knot!

To make a more extensive study of knots, mathematicians had to develop a different notation—an algebraic notation. This is analogous to the step René Descartes took in the seventeenth century when he showed how to reduce geometry to algebra.

In the 1920s, a mathematician called J. W. Alexander showed how to use algebra to describe the way a knot winds around itself, crossing over and under. This was quite a new way of using algebra, and it led to major advances in our understanding of knots. Since a computer can be programmed to handle algebra, it has also enabled mathematicians to use computers in their studies of knot patterns.

The discovery that knot theory can be applied to help the study of viruses is one of the most recent developments in the subject, beginning during the early 1980s. Mathematician De Witt Sumners and biologist Sylvia Spengler form one of a number of research partnerships that bring together mathematics and biology.

"Viruses have developed over the centuries to take advantage of other living systems," observes Spengler. "We have viruses that take advantage of plants, viruses that take advantage of bacteria, and viruses that take advantage of animals—horses, pigs, monkeys, humans, et cetera."

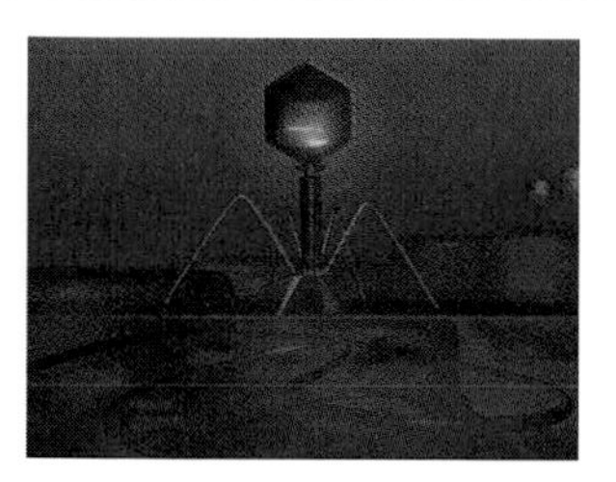

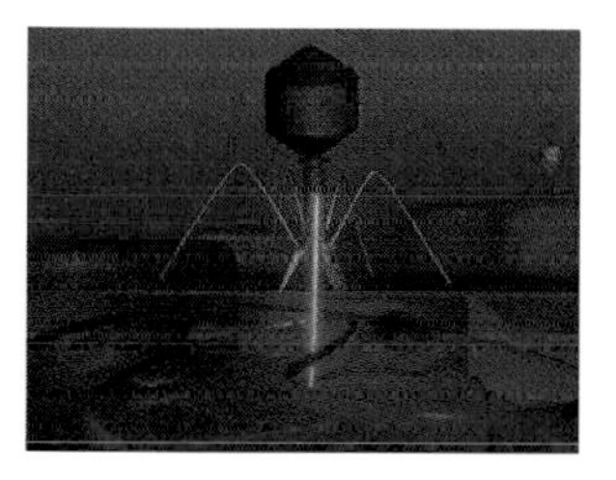

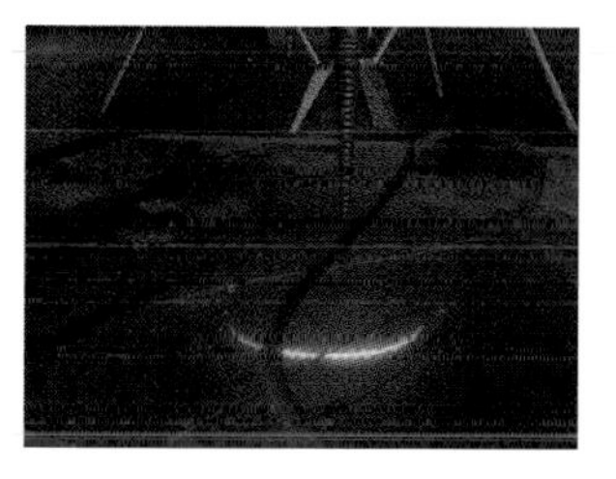

The computer animations at the left demonstrate how a virus invades and takes charge of a cell. In the first frame, the virus lands on a cell wall. In the second frame, the virus injects itself into the cell, penetrating into the cell's DNA. In the third and fourth frames, the virus fuses with the DNA. The virus then instructs the cell to create many copies of the virus, and in the last frame, the cell explodes, releasing the virus copies to go out and repeat the process in many more cells.

Though there are many different kinds of viruses, scientists start with a fortunate break, in that all viruses behave in roughly the same way. When a virus lands on a cell, it injects itself into the cell. Once inside the cell, it creates an enzyme that helps it to alter the cell's DNA, thereby enabling the virus to take control of the cell. After it has taken command, the virus instructs the cell to make copies of the virus, until the cell explodes, spreading the virus copies onto adjacent cells. In this way, the virus multiplies.

Spengler is trying to stop the virus from spreading by preventing the initial step where the virus inserts itself into the DNA. "If you can understand how the insertion step is carried out, and if you can identify the enzyme that does it for a particular virus, it gives you a target, a way of stopping the virus," she says.

DNA molecules are in the form of long, thin chains. In order to fit inside a cell, the DNA coils itself up. The enzyme causes the DNA to knot itself, thereby bringing close together sections of the molecule that the virus then swaps and amends, to create a quite different molecule. For example, one of the viruses Spengler studies ties the target DNA molecule into a trefoil knot, which crosses itself three times.

Away from the lab, Sumners uses techniques of pure mathematics to study the knots that

At the right is an electron microscope photograph of a DNA molecule tied into a knot by a virus. Below it is a drawing demonstrating how this DNA knot can be "unraveled" and identified as the double trefoil knot, which crosses itself seven times.

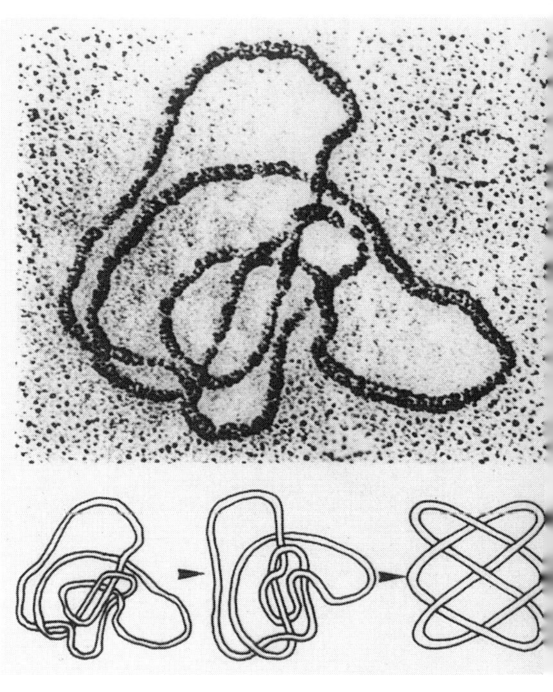

Spengler sees through the electron microscope. Dealing as they are with processes on a submicroscopic scale, the two researchers use the mathematics of knot theory to help make the invisible visible.

"Fighting viruses is a constant battle," Sumners says. "Any tools we can have in this battle are good ones. So it's 'Know thine enemy' in that sense. Anything and everything you know about them can and will be used against them."

These days, one of the weapons being used against viruses is the theory of knots. Over one hundred and fifty years since the subject was begun as a purely intellectual exercise, knot theory finds itself playing a major role in the fight against disease.

THE GEOMETRY OF FLOWERS

The different ways dinosaurs moved, the patterns of spots and stripes on the skins of animals, the way viruses cause DNA molecules to tie themselves into knots—from the large to the small, mathematics can help us to understand the world of living creatures. We can likewise use mathematics to help us understand the other living world, the world of plants.

For instance, how do you describe the shape of a flower? For a daisy, you could say it was circular. Thus, any mathematical description of a circle will give you a description of the shape of a daisy.

One way to describe a circle is as "the set of all points in a plane that are the same distance from a fixed point." (The "fixed point" is the center of the circle. The "same distance" is the radius of the circle.) This description corresponds to the way we draw a circle using a compass. The sharp point of the compass is placed on what will be the center of the circle. The compass point is sharp precisely so that the center is a "fixed point" that does not move around when we draw the circle. The compass is stiff so that once we set the angle of the hinge, the points traced out by the pencil are all the same distance from the center.

Another way to describe a circle is Descartes-style, using an algebraic equation.

But neither a daisy nor indeed any real flower is really circular. It only appears circular when viewed from a distance. When you take a proper look, you see that the flower is made up of many petals. The shape traced out by these petals is only approximately a circle; in reality it appears much more complicated than a circle. Can mathematics be used to describe the real shape of a daisy? Or how about a flower that is not circular, such as a lilac? Can mathematics describe the shape of a lilac flower?

The question seems to be pointless. After all, what possible benefit is to be gained from giving a mathematical description of a flower? Why would anyone bother to find out? The answer is that it is always a good idea to try to understand nature. Not only because, as human beings, we seem to take satisfaction from understanding our world, but also because we can never be sure when we might need our scientific understanding. After all, when mathematicians of the early nineteenth century first asked themselves, How can we describe the pattern of a knot? no one had any idea that biologists in the late twentieth century would use those mathematical descriptions to help in the fight against viruses. The lesson that history has taught us again and again is that scientific knowledge, which includes mathematical knowledge, generally turns out to be beneficial to us. Of course, it can also have harmful side effects, such as pollution from chemical factories, just as the benefits of having automobiles come at the price of road accidents. But on balance, the benefits far outweigh the costs.

The branches of a tree are a good example of self-similarity. If you look at the main branches growing out of the trunk, you will see that each looks like a smaller tree in itself, and so on as you look at smaller and smaller branches.

What might be a benefit of trying to find a mathematical description of, say, a lilac? Well, here is one possibility: it could lead to more accurate weather forecasts. Surprised? Here's how.

If you look closely at a lilac, you will notice that a small part of the lilac flower looks much the same as the entire flower. You see the same phe-

nomenon with certain other flowers, with some vegetables, such as broccoli or cauliflower, and with some other plants, such as ferns. Mathematicians refer to the phenomenon whereby a small part looks like the whole as "self-similarity."

What other things do we see that have self-similarity? Clouds, for one. If we had a mathematical way to describe self-similar patterns, we could use it to study clouds. With a good mathematical description of clouds, we could simulate the formation, growth, and movement of clouds on a computer. Using those simulations, maybe we could improve our ability to forecast severe weather, using our forecasts to protect ourselves better from the consequences of a major storm or a tornado. Fanciful? Not at all; researchers have been carrying out just such investigations for some years now. Nature may well turn out to be too unpredictable for us to ever have perfect weather forecasts, but the use of mathematics has already led to better forecasts, which almost certainly has saved lives.

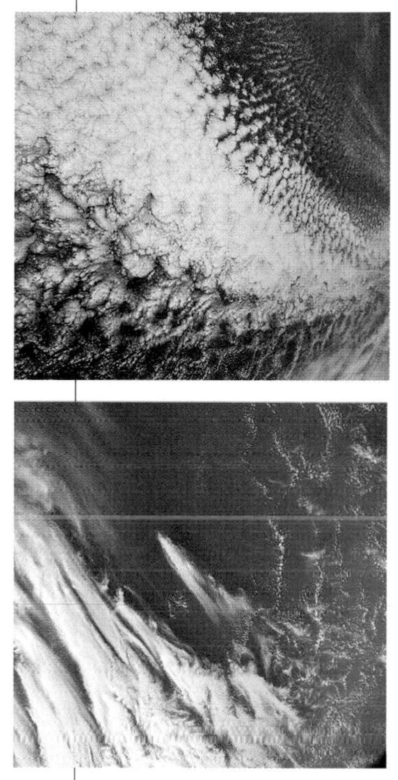

The two aerial photographs of clouds above reveal that different types of clouds have different characteristic shapes, and that clouds are made up of repeating patterns.

In other words, just as a mathematical study of the patterns of knots could lead to better techniques to conquer viruses, so too a mathematical study of the self-similar patterns of flowers such as the lilac can lead to better techniques to forecast the weather. This is the way mathematics works.

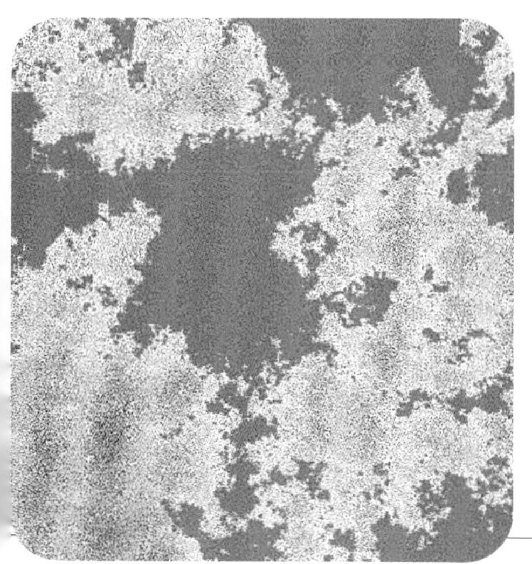

The computer simulation of clouds at the left was created by a fractal program and shows that scientists have come a long way toward creating the mathematics of clouds inside their computers.

Przemyslaw Prusinkiewicz is one of a number of researchers who have been trying to find mathematical descriptions of self-similar shapes such as the lilac flower. To obtain the description of a lilac, Prusinkiewicz and his collaborator, Dr. Campbell Davidson, look at the

"The beauty of nature can be appreciated from many angles. There is the visual aspect of it. Then there is the pleasure of understanding how things work."
PRZEMYSLAW PRUSINKIEWICZ
computer scientist

way nature might create the flower's shape. This is the same as arriving at a mathematical description of a circle by looking at how you can draw a circle with a compass. "When I am looking at plants," says Prusinkiewicz, "what I find beautiful is their shape and form. But there is another layer of beauty, a hidden beauty. It is not what we see at first sight when we look at a plant; it is the beauty of understanding the mechanisms that bring this form about."

Just as the mathematical theory of knots was developed long before biologists started to use it to study viruses, so too the mathematics Prusinkiewicz needs to study flowers began many years earlier, in this case at the end of the nineteenth century. One of the key discoveries was made by mathematician Niels Fabian Helge von Koch. He noticed that if you take an equilateral triangle, add a smaller

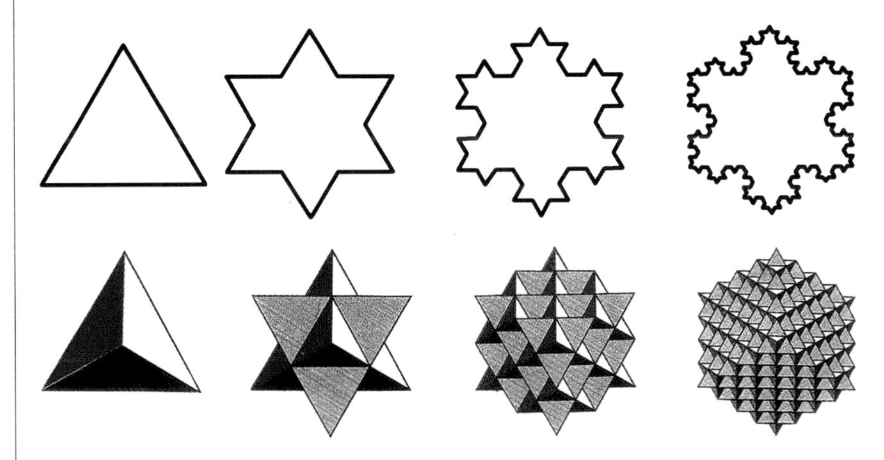

In the top row, a simple equilateral triangle develops into the traditional Koch snowflake image. In the bottom row, we see a three-dimensional representation of the Koch snowflake, which demonstrates the beauty of the symmetrical shape created through the repetition of the basic form.

equilateral triangle to the middle third of each side, then repeat the process of adding smaller and smaller triangles to the middle thirds of the sides, eventually you will develop a fascinating shape called the Koch snowflake. (To be precise, the idea is that you delete the middle-third length each time you add a new triangle.)

What the example of the Koch snowflake shows is that a complicated-looking shape can result from the repeated application of a very simple rule. It is the use of the same rule over and over again that results in the self-similarity of the resulting shape. Present-day mathematicians refer to self-similar figures as fractals, a name invented by the mathematician Benoit Mandelbrot in the 1970s. Mandelbrot showed that the repeated application of one particular kind of rule leads to a very important (to mathematicians) fractal shape that now bears his name, the Mandelbrot set. Computer images of the Mandelbrot set have shown that it is a figure of incredible beauty, and entire books and movies have been published that show nothing but different views of this one figure, at different degrees of "magnification."

Benoit Mandelbrot, the discoverer of fractals.

The famous "Mandelbrot set" fractal. As you look closely at the image, you will see more and more repetitions of the distinctive swirling "S" and beetlelike shapes.

A computer image of a lilac generated by Przemyslaw Prusinkiewicz with a simple mathematical rule.

For Prusinkiewicz, the idea then is to write down rules that, when used over and over again, produce the self-similar shapes he sees in nature, such as the shape of the lilac flower. Mathematicians refer to such a system of repeatable growth rules as an "L-system." The name comes from Aristid Lindenmayer, a biologist who, in 1968, developed a formal model for describing the development of plants on the cellular level.

For example, a very simple L-system to produce a treelike shape might say that if we start with the top part of any branch, that portion forms two new branches, giving three branches. When we repeat this rule on the new tops, we find we rapidly get a treelike shape.

To generate a lilac on his computer, Prusinkiewicz starts with a very simple L-system to generate the skeleton of the flower. Then, by taking careful measurements of an actual lilac, he refines his L-system so that the figure it produces more closely resembles reality. Using his refined L-system, he then generates the branching structure of the lilac. He then uses the same technique, with a different L-system, to produce the blossoms. And voilà! A lilac grows before his eyes. Not a real lilac, produced by nature, but a mathematical lilac, produced on a computer.

For Prusinkiewicz, it is a source of never-ending amazement to see how the seemingly complex shapes of nature can result from very simple rules. "It is very exciting to see that structures which we used to think of as very complex turn out to be very simple in principle," he remarks. "A plant is repeating the same thing over and over again. Since it is doing it in so many places, the plant winds up with a structure that looks complex to us. But it's not really complex; it's just intricate. When you appreciate the beauty of plant form, it comes not only from the static structure, but often also from the process that led to the structure. To a scientist who appreciates the beauty of this flower or leaf, it is an important aspect of understanding to know how these things were evolving over time. I call it the algorithmic beauty of plants. It's a little bit of hidden beauty."

Prusinkiewicz sees his work as creative. He uses mathematics to create (in his case, on a computer) some of the patterns we see in nature. "Creativity is the essence of mathematics," he says. "Mathematics is not playing with numbers and doing accounting. Mathematics is dealing with ideas in a creative and yet very precise way."

A particularly creative use of mathematics has been the development of artificial life—a simulated ecology in a computer.

THE JUNGLE INSIDE THE COMPUTER

Tom Ray has always been fascinated by the rain forests. For the past twenty years he has been studying them as a scientist. "The rain forest is full of life," he says. "Everywhere you turn there's some creature or plant or animal. There are so many different kinds, you'll never see them all. It's a vast unknown." To help him in his studies, Ray has built his own rain forest—inside a computer.

Whereas Prusinkiewicz is interested in understanding the shapes that nature creates, Ray wants to understand "the laws of the jungle." What are the patterns of an ecosystem, where different species vie or cooperate with one another for survival? Just like Prusinkiewicz, Ray uses mathematics to try to find an answer. For both researchers, the mathematical descriptions they obtain may not be exactly "the way nature does it." What both do, however, is show how simple rules can give rise to some of the complexities of nature.

As with knots and the fractal theory of clouds, both of which have proved to be of real benefit to mankind, so too Ray's mathematical models of ecosystems might turn out to be of great benefit in

The lush vegetation of a rain forest harbors an astonishing number of plant and animal species.

helping us to predict the effects on our environment of various decisions we make as inhabitants of the earth.

Ray began his computer experiment back in 1989. The idea was to create a virtual rain forest—an artificial ecosystem—inside a computer, using mathematics. He would populate his virtual world with artificial creatures he called digital organisms. Just as we can program a computer so that various parts of its memory represent numbers, which the computer can move to new locations in the memory or add or multiply together to form new numbers (for example, by doubling or halving), so too Ray set out to program his computer so that various parts of its memory would represent digital organisms. Like real living creatures, Ray's digital organisms could move around. He also formulated rules whereby a digital organism can reproduce. For example, one requirement might be that the organism has to be big enough—it should be an "adult." Another might be that there should be sufficient nutrients to sustain the life of another organism.

"Mathematics is not playing with numbers and doing accounting. Mathematics is dealing with ideas in a creative and yet very precise way."

PRZEMYSLAW PRUSINKIEWICZ
computer scientist

A further possibility that Ray wrote into his computer program was to allow for random mutations, where the offspring produced by a parent is not quite the same as the parent. This would prepare the way for evolution in his artificial ecosystem. According to the theory of (real) evolution, the occurrence of such random variations should lead to the emergence of completely new species.

Though the world Ray set out to create was completely artificial—a mathematical world created inside a computer—the aim was always to try to understand real life. "This project is about life," says Ray. "To me, evolution is the defining property of life and the creative force for the diversification of life. I think of evolution as being an artist. The artistic creations of evolution are incredibly beautiful and complex. Imagine a hummingbird pollinating a flower, or a cheetah running down its prey, or the human body. Those are fantastically beautiful things."

When he had finished putting his "rules of reproduction" into the computer, Ray breathed life into his artificial world by introducing just one organism. He called it "the Ancestor." Once the Ancestor was introduced, Ray simply sat back and observed what happened. He had programmed his computer so that different organisms would be displayed on the screen by lines of different lengths and colors. The result was not unlike the view of a sample of (real) pond life seen under a microscope.

What Ray saw that first night when he introduced the Ancestor surprised even him. "That first night it ran, I got pretty rich evolution," he recalls. "In a few hours I got hundreds of mutant versions of the original organism. The Ancestor evolved into a tree of digital life: a whole array of different species that are all connected, just like organic life on earth. Just as in real life, we had interactions between species cooperation, cheating, lying, and stealing."

"The purpose of computing is insight, not numbers."

R. W. HAMMING
pioneering computer scientist who worked on the Manhattan Project to construct the first atomic bomb during World War II

The digital organisms Ray sees evolving in his artificial computer world are analogs of the most basic forms of life on Earth: the bacteria, the algae, the protozoa, and the viruses. And yet it is from such simple life-forms that giraffes, cheetahs, corn, wheat, daisies, lilacs, and human beings have evolved. Ray sometimes wonders what would come of his artificial world if he were to leave it running for a hundred years or more. "If all we knew about digital organisms were the equivalent of algae and protozoa, I don't think we could fully envision the possibilities that are out there," he says. "It could be that evolution can reach a comparable level of complexity in the digital medium, in which case it would be truly fantastic in some future." The possibilities of this digital life in the future boggle the mind. As Ray says, "The excitement is about the potential. We're venturing into the unknown."

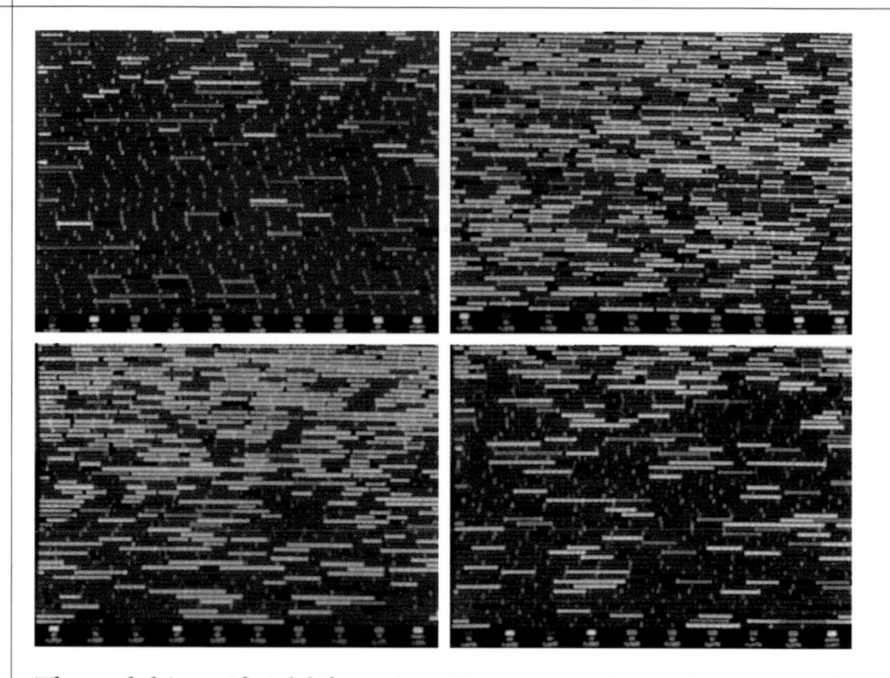

In these four screen images of Tom Ray's Tierra computer world, we see a survival-of-the-fittest drama played out. In the first frame, we see a population dominated by red creatures, represented by red line segments, along with a few blue creatures, into which some yellow parasites, represented by the yellow line segments, have started to invade. In the second frame, the population of red creatures has been severely diminished by the parasites. In the third frame, we begin to see the blue creatures, which are immune to the parasites, increasing in number and forcing the parasites toward the top of the screen. In the last frame the blue creatures have largely forced the parasites out and are now the dominant creatures.

Through his artificial-life project, Ray uses mathematics to provide yet another opportunity to make the invisible visible. Real evolution takes far too long to study, often millions of years for noticeable changes to occur. In Ray's computer, the same kinds of changes can happen overnight, and the next morning he can view the "highlights" as an action replay.

Tom Ray's work, modeling evolution, is in many ways similar to that of computer scientist Przemyslaw Prusinkiewicz, who models the growth of flowers. Prusinkiewicz's models are based on L-systems, mathematical rules that capture the cellular growth patterns of real plants. Ray's models are based on our knowledge of how evolution occurs. Prusinkiewicz sums up the modeling approach to science in these words: "The process of modeling is quite inseparable from the process of learning about nature. What a model is, is an assembly of those features of nature which we believe are important. By creating the model, you can verify whether indeed all those aspects are important, and whether you haven't missed anything."

With his artificial world, where digital species evolve and grow inside a computer's memory, Ray joins the many others who have used mathematics to help understand the patterns of life—from huge dinosaurs to the microscopic viruses that invade our bodies, from the skin patterns of animals to the shapes of flowers, from the extinct past to the present day. And with his models of evolution, perhaps Ray can even give us a mathematical glimpse into the future, by letting us see the patterns by which evolution works its changes.

Chapter 4

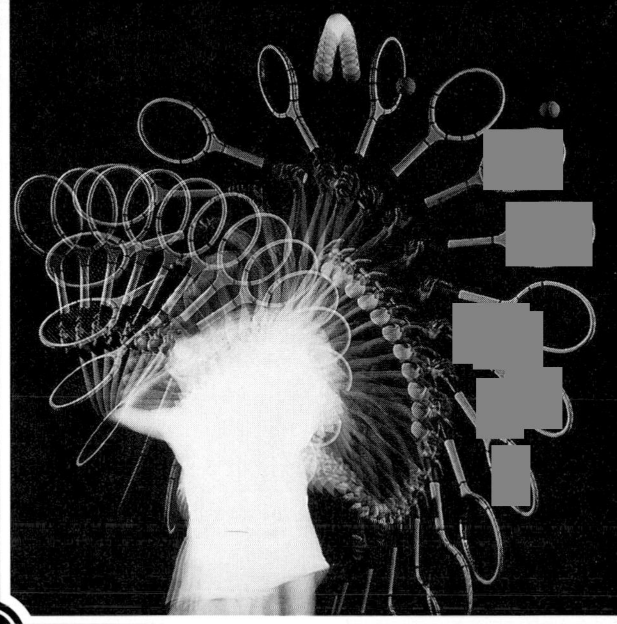

THE NUMBERS GAME

Since the time of the ancient Greeks, athletics has been regarded as one of the highest forms of human expression. For many, athletic competition brings out the best in us and expresses the epitome of the human potential.

What does it take to do well in athletics? What is required in order to become a top performer— to reach the pinnacle of your chosen event?

Everyone knows that to succeed in any sport, you need talent, skill, and a whole lot of training. These days, increasingly, athletes and coaches are turning to something else as well: mathematics.

This mosaic from ancient Sicily depicting women exercising in a gym attests to the antiquity of the ideal of physical fitness.

They have realized that the application of mathematics to the different human endeavors we call athletics not only enables us to appreciate more fully the drama and potential of this human expression, but also holds a key to unlocking the hidden potential deep inside us all. Mathematics can point the way to increased performance, and can help find the crucial edge that marks the difference between winning and losing.

BALLS IN THE AIR

The tennis player completes an elegant backhand return. The golfer performs a perfect swing. The baseball pitcher releases the ball from his hand with a final flick of the wrist. What happens next? What happens to the ball once it has left the racket, club face, or pitcher's hand? What is the effect of topspin on the flight of the airborne tennis ball? How do the dimples in the surface of the golf ball affect its motion through the air? Is there really such a thing as a breaking curveball?

The only way to find reliable answers to questions such as these is by mathematics. During the past twenty years, scientists and athletes have been using mathematics to better understand the mysteries of how balls of all shapes and sizes move through the air. The answers were to be found in the science of aerodynamics, using the same mathematics used to design aircraft—mathematics that is over 250 years old.

THE SECRET OF FLIGHT

Few air travelers, sitting in an aircraft seat at thirty thousand feet, sipping a martini, are likely to pause to ask themselves what is keeping them in the air. The technical answer is, the wings. The scientific answer is, an equation. Bernoulli's equation, to be precise.

Daniel Bernoulli was one of a large family of unbelievably talented and productive Swiss mathematicians who lived in the eighteenth century. Daniel's father, Johann, was a professor of mathematics at the University of Basel. Both father and son were highly influenced by the method of infinitesimal calculus invented in the previous century by Isaac Newton in England and Gottfried Leibniz in Germany, and both had helped to develop the new technique.

Left, Johann Bernoulli.

Below, Daniel Bernoulli.

Calculus provides a way for the mathematician to study motion. Prior to the middle of the seventeenth century, mathematics could really only be applied to static things. Mathematicians could count, they could measure length and angle, they could study shape, they could calculate slopes of straight lines, areas of figures, volumes of solid objects. But they could not write down equations that described the motion of a ball thrown into the air or a planet as it orbits around the sun.

The problem of how to describe the motion of planets particularly fascinated the young Isaac Newton. To solve the problem, he invented an entire new branch of mathematics: calculus. Calculus allows you to use the static techniques of mathematics in order to study moving objects. It is similar to making a movie.

These days, everyone knows that if you take a sufficiently rapid sequence of still photographs of a moving scene and project them onto a screen—thirty shots a second or faster will suffice—then the human eye will see the result as continuous motion. The small differences between one frame and the next cannot be detected by the human visual system. Newton's idea was likewise to regard continuous motion as made up of a sequence of still configurations.

Each still configuration could be analyzed using existing mathematical techniques—principally geometry and algebra. The difficult part was putting all the still configurations together. Thirty frames a second will fool the human brain into thinking it is seeing continuous motion. To achieve the same result—continuous motion—mathematically, Newton had to "project" his still configurations at infinite speed, and each still configuration had to be infinitely short in duration. What we nowadays call "calculus" (more accurately, "infinitesimal calculus") is the collection of techniques Newton (and, later, others) developed to perform this "infinite sequencing."

Sir Isaac Newton.

For Newton, the goal was to use his new methods to study the motion of planets. But what could be done with planets could likewise be done for any moving objects—cannonballs, for example. Daniel Bernoulli wondered if the same ideas could be adapted to study the continuous motion of fluids (which to a scientist means liquids or gases). On the face of it, this was a very different problem.

It took Bernoulli most of his adult life, but he did it. In 1738 he published his results in his book *Hydrodynamics*. In that book, Bernoulli derived the equation that now bears his name, the equation that forms the basis of aircraft flight. In essence, what Bernoulli's equation tells you is that when a fluid flows over a surface—such as an aircraft wing—the pressure the fluid exerts on the surface decreases as the speed of flow increases.

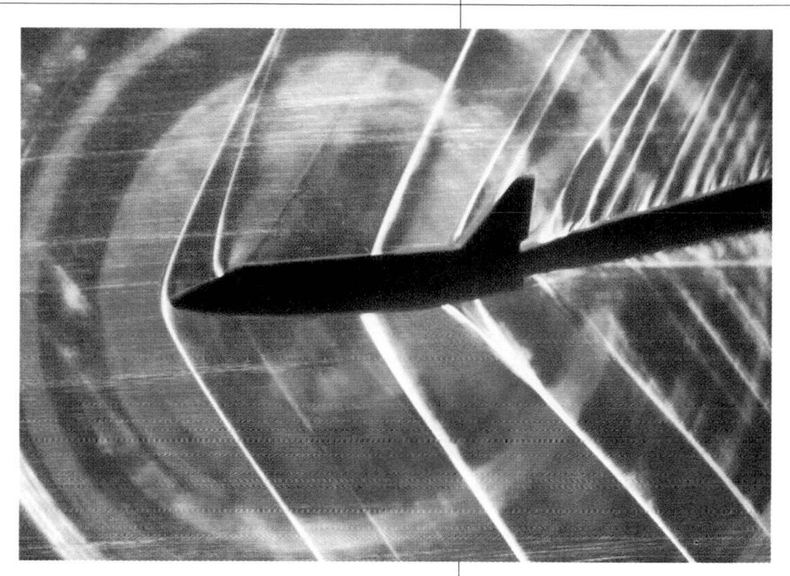

This image of a wind-tunnel test of an airplane shows the air-current patterns created as the plane cuts through the air.

That simple relationship is the secret of flight. An aircraft wing is shaped so that it is curved on the top, flat on the bottom. As the aircraft is propelled forward, the wing splits the air in two: the air that travels over the top of the wing and the air that travels underneath the wing. Because of the shape of the wing, the air has to travel farther around the curved top than along the flat bottom. To do that, it has to travel faster. As a result, the air pressure exerted on the top of the wing is less than that exerted on the bottom. Voilà: the air pushes the wing—and with it the aircraft—upward. Aircraft engineers call it "lift." The faster the plane moves forward, the greater the lift. (That's why airplanes need to gather speed along a runway in order to take off.)

"But what about small fighter aircraft that can fly upside down, or those early biplanes that had thin, flat wings?" you might cry. The above explanation using Bernoulli's law does not seem to apply in those cases. So what keeps those planes up? The answer is that there is a second source of lift.

For a modern jet airliner, the Bernoulli effect caused by the shape of the wing is what keeps the plane aloft in level flight. But you can also get lift from the "tilt" of the wing, the angle the wing makes with the onrushing airflow. Even at fairly slow speeds, a wing that tilts upward at the front will create lift, as the tilted wing compresses the air in its path. The flat wings of those early biplanes slanted upward, which gave them their lift, and modern jet airliners point their noses upward very noticeably during takeoff to obtain additional lift.

It is so often the case in mathematics that many years elapse between a mathematical discovery and its application. In the case of Bernoulli's equation, it was to be over a hundred and fifty years before the days of powered flight, and two hundred years before airplane wings were designed scientifically to take the maximum advantage of the lift guaranteed by the mathematics.

The first golf ball was a leather sack full of feathers, called the Feathery, circa 1400.

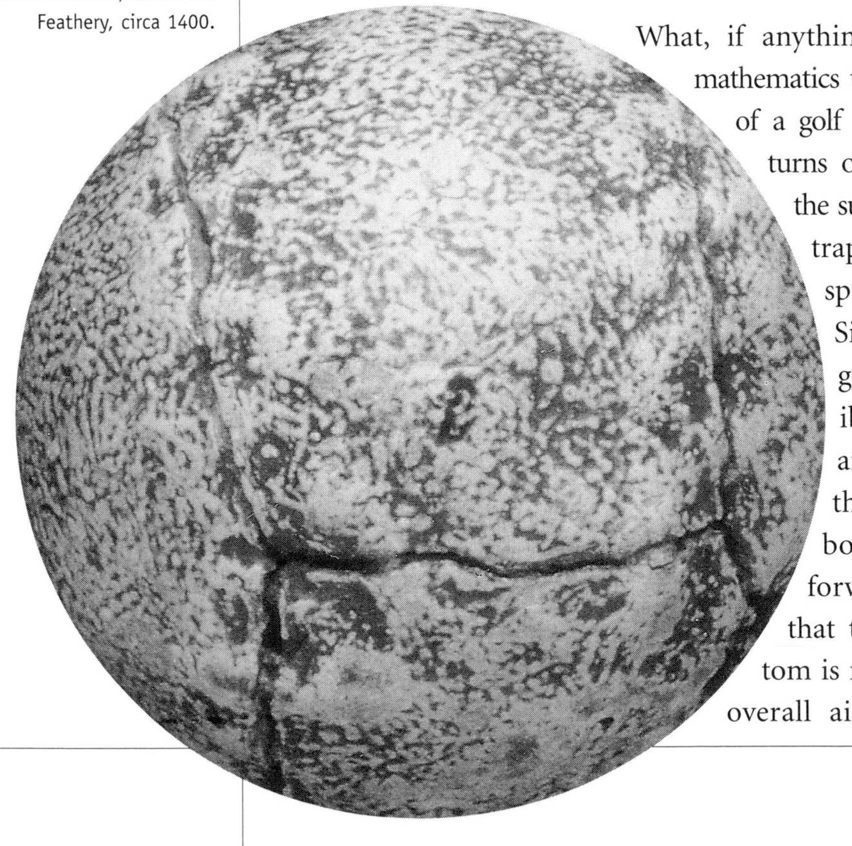

What, if anything, does Bernoulli's mathematics tell us about the lift of a golf ball in flight? As it turns out, the dimples on the surface of a golf ball trap a layer of air that spins with the ball. Since a correctly hit golf ball has incredible backspin, the air layer trapped by the dimples at the bottom is pushed forward. This means that the air at the bottom is moving against the overall airflow around the ball, which is from front to back, as the ball flies through the air. Similarly, the layer trapped at the top is being pulled backward, going in the same direction as the overall airflow. Thus, the top layer is going faster than the bottom layer. (Imagine running up and down an escalator. When you run in the same direction as the escalator, your overall speed is faster; when you run in the opposite direction, your overall speed is slower.)

By the late nineteenth century, golfers had discovered that a ball's flight was longer and more accurate after it had been scarred by repeated use. Ball manufacturers began to develop balls to take advantage of these effects, as in the popular model shown on the left, the Bramble, circa 1899.

By Bernoulli's law, the difference in airspeed caused by the backspin on the golf ball means that the pressure on the bottom of the ball is greater than the pressure on top. In other words, the spin of the ball creates lift, just as the airflow over an aircraft wing creates the lift that keeps the aircraft in flight. It is the lift created by the dimples that enables a golf ball to travel as far as it does—a smooth golf ball would travel less than a quarter of the distance of a standard ball.

Today, golf balls are made with a dimpled surface because manufacturers have discovered that dimples give the ball greater air lift.

Bernoulli's two hundred and fifty year old equation also resolved a question that had puzzled generations of baseball fans: Can a baseball pitcher throw a breaking curveball, a ball that curves in flight and then suddenly drops before it gets to the plate? The players said yes, they saw it all the time. The scientists said no, it was impossible.

A mathematical analysis showed that both were right. Though a breaking curveball is indeed a physical impossibility, batters and catchers were seeing something that looked like a sudden drop in the ball. The explanation was a mix of Bernoulli's equation and perspective. Here is what happens.

A baseball has 216 stitches on its surface. As the ball spins in flight, the stitches pull a layer of air around it, just as do the dimples on a

The much disputed "breaking curveball" is an optical illusion resulting from the effect of sidespin on the ball's path.

golf ball. If the baseball is thrown sidespin, the air on one side flows faster than on the other. According to Bernoulli's equation, this creates higher pressure on one side, pushing the ball in a curve. Similarly, by imparting topspin to the ball, the pitcher can cause the ball to curve downward. The question was, could topspin develop over the life of the pitch to cause the ball to fall suddenly?

The physicists said no, and they were right. Slow-motion film of baseballs in flight showed that the ball traveled in a consistent curve. If the force of gravity were eliminated, and if the catcher were not in the way, the ball would move in a perfect circle and return to the pitcher. (It would in fact hit him in the back of the head.) For the hitter standing near the perimeter of the circle formed by the curving ball, the ball appears to be moving in a straight line at first. However, as it approaches, its path appears more and more vertical—thus its apparent breaking motion. From their perspective, the batter and the catcher are deceived into "seeing" a sudden drop in the ball.

From golf to baseball, the same eighteenth-century mathematics can be used to understand, to explain, and, sometimes, to improve performance.

THE TECHNOLOGY OF IMPACT

We read a lot about the impact of technology. But what about the technology of impact? For example, what exactly happens when the strings of a tennis racket come into contact with the ball? Attempts to answer this question have led to significant changes in the way tennis rackets are designed and built. Those changes in equipment

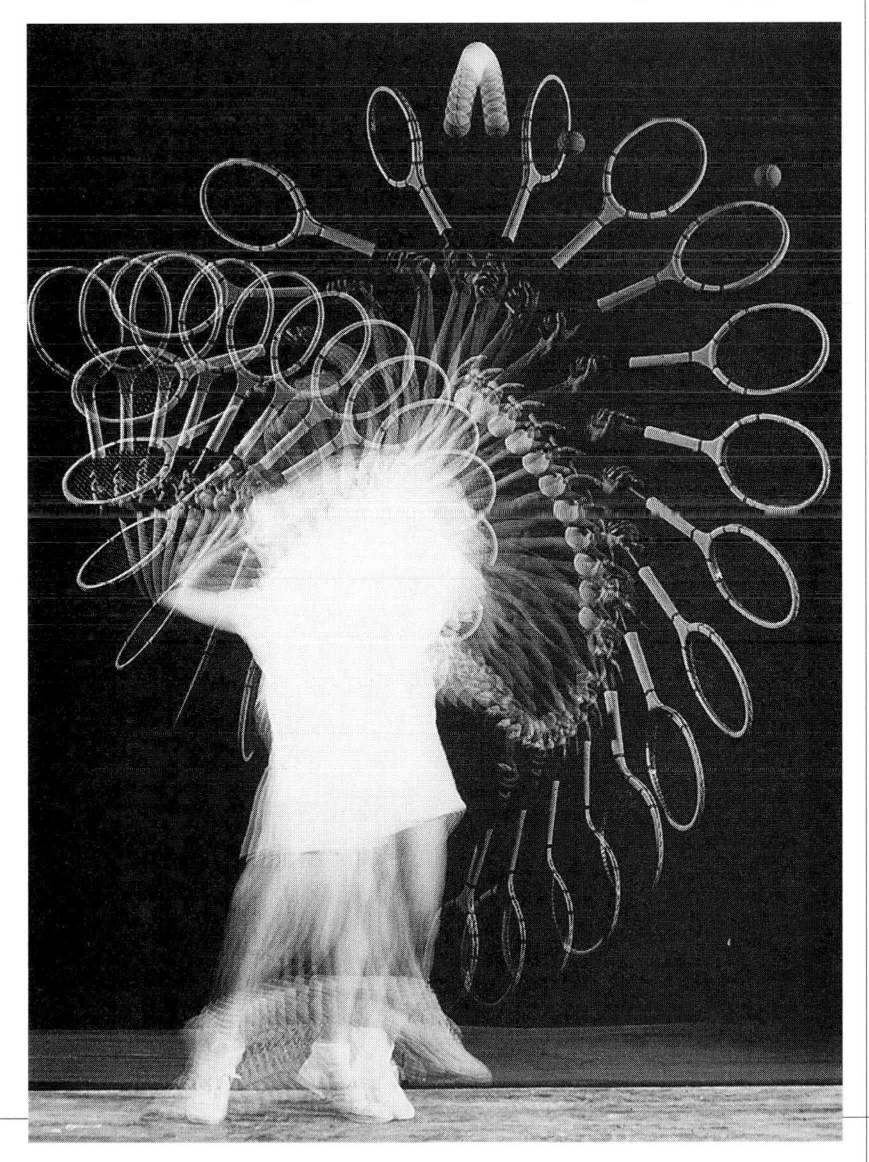

Harold Edgerton's strobe photography analysis of a tennis serve.

have led to changes in the way the game is played, resulting in a much faster game than in the past. Like most changes in technology, they have used mathematics.

Aerospace engineer (and tennis player) Rudrapatna Ramnath is one of the scientists who have been at the leading edge of recent developments in impact technology. "I've played tennis all my life," he reflects. "There have been changes that are remarkable to witness. The ball has actually changed very little. What has changed is the racket." For over twenty years, Ramnath has measured and analyzed tennis rackets in his laboratory at MIT. "The purpose of the measurements is to accurately evaluate performance. The challenge has been to create a systematic approach to evaluating a racket, measuring its performance in a consistent numerical form."

"Defining the elements of a game in mathematical terms allows you to model the game and predict how changes will affect it."
RUDRAPATNA RAMNATH
aerospace engineer

Ramnath's analyses began with high-speed photographs of rackets hitting tennis balls. By using very fast film in high-speed cameras and lighting the scene for very brief instants by a rapidly flashing light source known as strobe lighting, it was possible to capture on film a sequence of frozen moments as the racket comes into contact with the ball. The technique is called "strobe photography." It was developed by Harold Edgerton shortly after the Second World War to provide scientists (and, more recently, artists and advertisers) with a way of revealing things that moved too fast to see with the human eye.

For example, strobe photography has produced photographs that show what happens when a drop of milk splashes and what happens when a bullet passes through an egg. Recently, scientists have used the technique to find out what happens when you prick a soap bubble with a pin. Does the entire film disintegrate at once, or does it break apart gradually, starting from the site of the pinprick and spreading across the whole bubble? (See the pictures at the right for the answer.)

By means of strobe photography and newer techniques using radar and infra-red technology, Ramnath makes precise measurements of the effects on the ball of the impact of the racket. (To focus the camera or the radar device, he holds the racket fixed in a clamp and shoots balls at it from a gun.) Using other instruments, he measures such features as the mass distribution, the damping, the stiffness, and the vibration. By taking measurements with different rackets, including experimental ones, he investigates the effects of different racket geometries, different string materials, and different string tensions.

Over the years, Ramnath and his students have measured over two thousand rackets. His results have been published in the magazine *World Tennis,* helping readers to choose from the growing array of rackets on the market.

Starting with his data, Ramnath has created mathematical models that help predict how a racket with specific features might perform. "We've learned a lot over the years," he says. "As technology gets more sophisticated, rackets have become more efficient." As a result, Ramnath foresees the day when he, or someone like him, is asked to provide mathematical rules to limit the kinds of equipment players can use: "As rackets get more and more efficient, many people feel the game's interest level is beginning to drop. Speed isn't everything. For the men's game, it's becoming a serving duel. These balls are moving faster than a hundred and twenty-five miles per hour and are impossible to return. It gets dull after a while, and audiences have started to complain.

These strobe-photography stills of a soap bubble bursting clearly show that a bubble does not disintegrate all at once.

"After measuring racket power for twenty years," he continues, "my job now is to create an upper limit for that power. This will help establish a rule that tennis bodies can use to ban rackets exceeding the limit. The measuring process is unchanged, but the new challenge is to define in mathematical terms a racket that will preserve the qualities that people love about the game."

A similar change took place in professional baseball in 1969. To eliminate the growing dominance of fast pitchers, the mound was lowered from sixteen inches to ten inches, and the strike zone was dropped from the batter's shoulders to the armpits and from the bottom of the knees to the top. A statistical analysis of the effects of the changes showed an increase in batting averages of seven points in the National League and sixteen points in the American League.

"Rewriting the rules of a sport is not something to take lightly," Ramnath acknowledges. "But it can be done with care and precision using the language of mathematics. Defining the elements of a game in mathematical terms allows you to model the game and predict how changes will affect it. With statistical analysis, mathematics can be used to evaluate the results to see if the predictions were correct."

SAILING FASTER WITH MATHEMATICS

Ocean racing is another activity where the mathematics of aerospace and modern technology find their way into the design of sporting equipment—in this case, large racing yachts.

The America's Cup is the premier event in ocean sailing. Competition is fierce. The technical challenges are enormous. And the costs are huge. These days, competitors look to mathematics to provide the crucial innovation that can mark the difference between winning and losing.

The America's Cup originated in 1851 when the boat named *America* won the All National's Race at Cowes, Isle of Wight, competing against fourteen British yachts for the trophy then known as the "Hundred Guineas Cup." The *America* is represented here by an unknown painter as she crosses the finish line. The boat became a popular symbol of American engineering achievement.

Traditionally, American groups called syndicates fight to represent the United States against foreign competition. For almost a hundred years, American boats successfully defended the cup. But recently, Australia and New Zealand have taken the prize. To win the America's Cup in 1991, one syndicate spent close to a hundred million dollars. Most of the money was spent on the design of the boat.

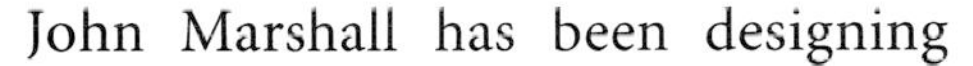

John Marshall has been designing America's Cup boats for over twenty years. "To cross the line first," says Marshall, "a boat must be skippered by a strategic genius, and the crew must be honed to a finely tuned machine. But the best skipper and crew can't compete successfully without a winning boat."

In 1995, Marshall headed the syndicate whose boat went up against New Zealand. They lost. He now runs a syndicate called PACT 2000 (Partnership for America's Cup Technologies 2000) that is trying to win it back. In their search for the winning boat, Marshall and his design team rely on computer simulations and mathematics. He says, "In this day and age, it costs too much and the design problems are too complex to build a lot of boats and test them in the water. Boats are modeled and tested on a computer long before we start pouring the fiberglass. Basically, this sport would not be possible today without mathematics."

Not only is mathematics at the heart of America's Cup racing when it comes to boat design, but the race may be the only sport that is actually defined by a mathematical formula. In order to ensure fair competition, the International Racing Committee has established a rule that limits the sail area and hull size (or displacement). The

Young America out in front during a race of the 1995 America's Cup.

bigger the hull, the smaller the allowable sail area, and vice versa. The rule is stated in the form of a mathematical formula. The goal facing the designer of an America's Cup boat is to find the optimal relation among the three key parameters: sail area, hull size, and keel.

Some of the problems facing Marshall are much the same as the ones mathematicians encounter when trying to understand the motion of balls through the air. For Marshall, though, the situation is more complex. An America's Cup boat has a much more complex shape than a ball, and it moves through two media at once, air and water. It's Bernoulli's mathematics of fluid flow with a vengeance.

"This sport [America's Cup racing] would not be possible today without mathematics."
JOHN MARSHALL
America's Cup racer

Marshall attacks the design problem by breaking it down into parts: the sails, which harness the wind to propel the boat; the hull, which must move smoothly and efficiently through the water; and the keel, which gives the boat stability. The weight and shape of these three elements, and the way they interact, will determine the speed of the boat. Marshall looks for minute increases in efficiency. "A one percent increase in speed may not seem like much," he admits. "But that translates into an eight-minute advantage in races that are often won by seconds. So the difference between winning and losing physics is not very much. That's why design is so crucial.

"The difficulty is trying to optimize a problem that has many parameters that trade off with each other. For instance, you don't simply take all the sail area you can get, because you have to pay for it in some way. That kind of optimization problem is not unique to sailboat design. It's common to economics problems, to business management problems, to a vast array of real-life problems. What you do is construct a mathematical model that includes all the important variables and all the equations that relate them.

"Take the width of the boat, for example. We can write equations that relate the width of the boat to the stability of the boat. We can also write equations that relate the stability of the boat to its performance in varying speeds of winds—essentially unimportant in very light wind, critically important in heavy wind. So now you have a relation between the width of the boat and a performance parameter which in turn is related to wind speed. Then we go back to the width of the boat and look at its other effects on performance, say, the wetted surface or frictional drag.

"So, gradually you build up a series of equations that are all interlinked, and which describe the entire physical system. A designer would now be able to select a set of parameters for the boat, choosing a number for each one, a length, a weight, sail area, and so on, and get a quantitative prediction of performance."

"As we move forward into the next century, the fundamental literacy that's required is the literacy of science—to understand that there are physical laws that can be expressed mathematically, and that there's a way of making decisions for society based on some analysis, rather than a purely emotional approach."

JOHN MARSHALL
America's Cup racer

In designing a boat, Marshall has to take account of the weather conditions that are likely to prevail on the day of the race. This means that the computer simulations have to include wind and wave models. Getting it right isn't easy. In 1995 in San Diego, everyone predicted light winds and small waves. But at race time, the winds were strong and the water was choppy. Australia's boat wasn't built to take the stress and it fell apart and sank. The winning boat was designed for the rougher conditions that prevailed.

With wind, wave motion, hull shape, and sails modeled by equations in the computer, Marshall and his team of designers display the movement of the hull through the water on a computer screen in the form of schematic diagrams. By changing the values of certain coefficients in the equations, they can study the performance characteristics of hulls of different shapes and sizes. In 1995, they examined twenty-three different computer models. Of those, they selected five for which they built scale models to test both in a water tank and in a wind tunnel. "The mathematics of the computer model must be translated precisely to the model for the tests to mean something," Marshall explains. "We're talking about tolerances of less than a thousandth of an inch."

"A lot of the mathematics and technology we used to create the boats was developed in the Cold War to create weapons. It's very sophisticated stuff."
JOHN MARSHALL
America's Cup racer

The data from the tests on the physical models were then entered back into the computer, enabling the designers to test the five designs against one another in virtual races held in different conditions. The winning design from these virtual races was then chosen for the real boat that was actually built and raced in the trials to decide who would represent the United States in the America's Cup. It was called *Young America.*

In the trials, *Young America* competed against *Mighty Mary* and *Stars and Stripes*. Even during the trials, mathematics continued to play a major role. Performance data were continually sent by radio to the chase boats, where computers were used to compare the actual performance with the mathematical prediction, allowing fine adjustments to be made.

Looking back over the whole design and testing process, Marshall comments: "It would be impossible to design a boat with a group of people working on it if there weren't some common language, some common vocabulary. Essentially, the mathematics of yacht design is the way to produce a matrix of better versus worse, a matrix of progress, an agreement among the team members that progress is

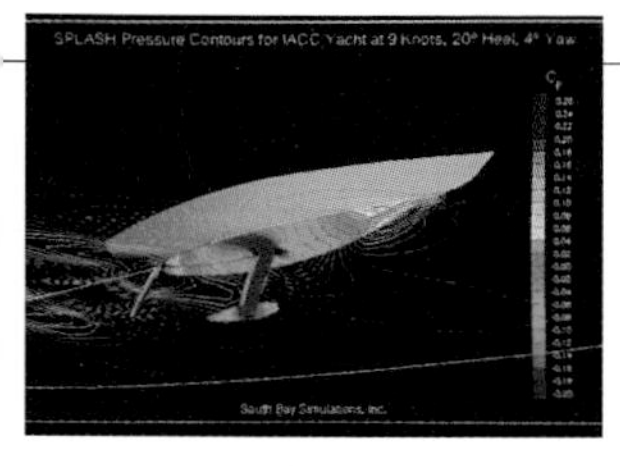

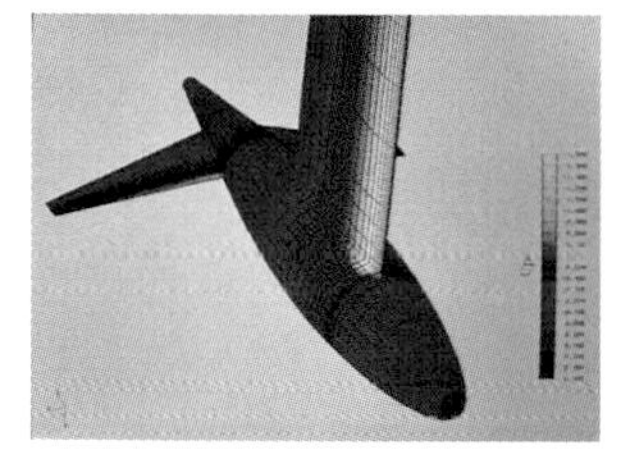

Left, computer simulations of the patterns of pressure around the hull and the hull appendage of a yacht. Simulations such as these allow engineers to test different designs for boats before going to the expense of having a model built.

taking place, and a way to integrate the ideas from all the team members into the final product. The mathematical answer is trusted as being the best available—not necessarily the correct answer, but the best available answer."

In the end, Marshall and his team did not win. Dennis Connor's *Stars and Stripes* was chosen to represent the United States against New Zealand. Marshall admits his disappointment: "It was difficult to lose after all that effort by so many people." He adds: "But I believe we did have the fastest boat." His opponents agreed, and when the America's Cup was run, Connor actually used Marshall's boat instead of his own. But in the rougher water that prevailed on the day of the race, the stronger New Zealand boat won.

Disappointed but undeterred, Marshall and his team went back to the lab to start designing what they are sure will be the winning boat in the year 2000. As Marshall remarks, "Winning sailboat races by a few seconds is a more stringent demand than the fuel efficiencies Boeing strives for." Among the many uncertainties of ocean racing, one thing is certain: if the PACT 2000 team is not successful, it will almost certainly not be because of having inferior mathematics.

A scale model of a hull being tested in a water tank.

TO SPIN OR TO JUMP?

Mathematics, graphs, charts, computer models—the scientific approach it takes to win the America's Cup is also what you need to win an Olympic medal in figure skating these days. As in most other sports, the use of mathematics is a very recent development.

In the 1980s, figure skaters from Eastern Europe introduced into international competition a maneuver known as the triple axel. As a result, they dominated the world figure-skating scene, taking all the medals. There was clearly only one way for American skaters to get back into the competition: they would have to master the technique themselves.

A triple axel requires the skater to launch into the air from one foot facing forward and perform three complete rotations in midair before landing smoothly and continuing to skate forward. Prior to 1980, it was considered a daredevil maneuver, likely to go wrong, not something to be attempted in a major competition. Though some American skaters were able to perform the maneuver, no one really knew how best to do it, and the results in competition were not impressive. A major problem was that the coaches did not know how to teach it. Was the secret to jump high or to spin faster?

Opposite (left) is a model of the triple axel (top), the double axel (center), and the single axel (bottom). The comparison of these three models showed that the skater attains the same height for each jump and that the most important difference in the jumps is how tightly the skater draws his or her arms in to the body, allowing for faster rotation.

U.S. Olympic figure-skating coach Kathy Casey needed to know that answer if her skaters were going to be able to win any medals. "Scott Hamilton in 1984 was the last skater to win the Olympics without a triple axel," she remarks. "By 1988, they were relatively common. Without them now, you're toast." To find the answer, Casey turned to the relatively new science of biomechanics—a science that, like the triple axel, originated in Eastern Europe.

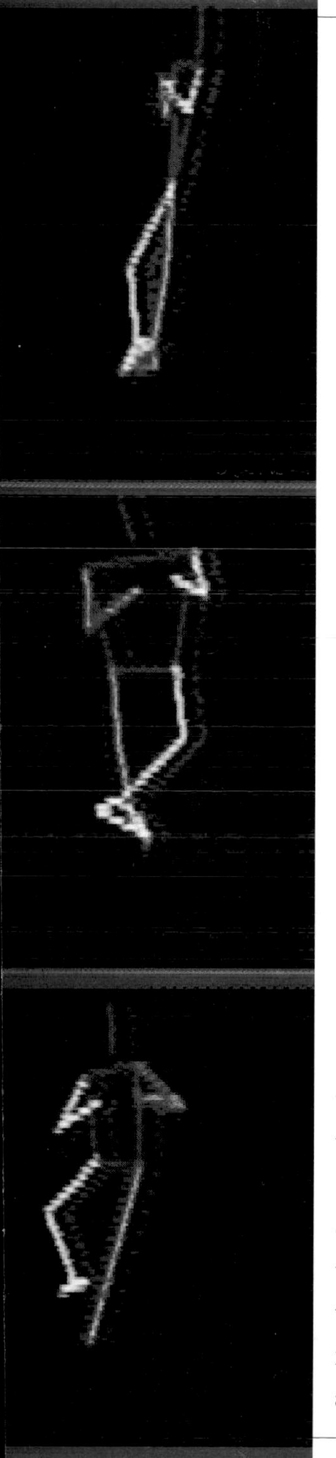

Biomechanics researchers at Pennsylvania State University set about analyzing the triple axel. The first step was to translate the maneuver into the language of mathematics and simulate it on a computer.

Debbie King was one of the team members who performed the work. For King, biomechanics was the perfect career. As a student, she majored in mathematics and biology, and she loves sports. Together with her colleagues, King set up three video cameras to capture skaters performing double and triple axels. King picks up the story: "Kathy Casey wanted to know how the physics changed from the double to the triple. The limiting force, of course, is gravity. The question was whether you had to spin faster or jump higher to do the triple."

Each camera showed the move from a different angle. King marked the skater's key joints in each video frame. Using techniques from geometry, she was able to combine the information from the three cameras to translate her data into a three-dimensional mathematical representation of the jump. King explains the next step, which she carries out using the computer: "Every sixtieth of a second I have the x, y, z coordinates of each joint in space. By measuring the distance between joints from one frame to the next, I can compute the speed that joint is moving as it progresses through the move. I can also measure joint angles precisely."

To create computer models of skaters performing the triple axel, Debbie King applied white dot markings to the clothes of the skaters and then videotaped them jumping, as shown above in this picture of Damon Allen. By plotting the white dots through the course of the jump, she was able to reconstruct the exact trajectory of the jump and to create a stick-figure representation in her computer of the skater performing the jump. These computer models of each skater's jumps allowed her to make precise comparisons between good jumps and bad ones.

King and her colleagues carried out the study on seventeen skaters. By comparing good jumps to bad ones, they were able to derive the optimal launch speed of the leading foot, an important question for coach Casey. They were also able to confirm Casey's suspicion that the key to the jump was to spin faster, not to jump higher. "The analysis shows the time of the double and triple jump is the same," Casey explains. "To do three instead of two spins, the skater must spin faster, putting all energy into a rotational motion. Though jumping higher would increase air time, it wastes energy. The objective is now clear."

As a result of the earlier research, these days Casey is able to teach young skaters how to perform the difficult maneuvers they will need to master in order to win in international competition. Damon Allen is one of those young hopefuls. "Damon has real potential," Casey remarks, "what we used to call innate talent. The coach's job is to nurture that talent."

"Before the study, we believed that you need to throw your leg out further for a triple," comments Damon. "After the study, we learned that in reality it needs to be closer so you can get into the air and into your rotation much quicker and get the job done sooner."

Watching Damon practice his triple axel, biomechanist King comments: "This complex movement takes less than a second to perform. It becomes more amazing to me to look beneath the surface of it and see what the body is actually doing in this brief moment. It's rewarding to give Kathy some good information for her and Damon to work with."

Casey has no doubts about the benefits she gained from the mathematical analysis: "Great coaching and good intentions are not worth anything unless the goal is correct. The mathematical analysis has helped me focus my work with Damon. It's like turning on a light in a dark room. We're not groping around. We know what we need to do."

WORKING THE SYSTEM

These days, when he trains, athlete Tim Deboom competes against an unusual opponent: a mathematical model of himself, based on his own earlier performances.

Deboom is one of a growing number of world-class athletes for whom the new science of biomechanics has transformed the way they prepare for competition. His specialty is the triathlon, a new Olympic event, in which the competitors complete a 1,500-meter swim, a 40-kilometer bicycle race, and a 10-kilometer run, one right after the other. Deboom is being trained by U.S. Olympic coach George Dallum, at the Olympic Training Center in Colorado Springs.

Triathletes swim neck to neck in the grueling swimming portion of their three-leg competition.

Dallum works out a unique training program for each athlete, based on the data obtained from a battery of tests performed on the individual. In Deboom's initial testing phase, data are obtained for all three events, over a three-day period.

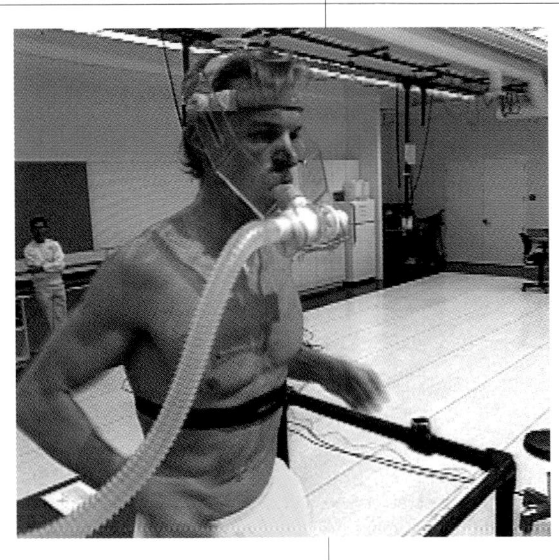

Tim Deboom hooked up to measuring apparatus on the treadmill.

Day 1: Running. To obtain physiological data of Deboom running, Deboom takes to the treadmill while monitors record his heart rate and oxygen uptake. Blood tests taken at intervals measure the rate at which he produces lactic acid as he nears his maximum threshold of activity. Lactic acid levels are particularly important in endurance events such as Deboom's, since lactic acid is what causes stiffness and the feeling of tiredness in the muscles.

Our muscles produce lactic acid all the time, as a by-product of action, but generally the body clears it away as fast as it is produced. However, when the muscles are asked to perform at a high rate for a prolonged period, the body cannot clear the lactic acid away fast enough, and it builds up. The muscles start to feel tired. Eventually, if the lactic acid reaches too high a level, the athlete will not be able to complete the event successfully. From the physiological data, Dallum can work out a training schedule that will ensure Deboom improves without overstraining his body's capacity.

Videos of Deboom's running action enable Dallum's assistants to measure the angles of his joints at different stages of his motion. Using this data, coach Dallum can suggest minor changes in his action to utilize his energy more efficiently.

To study an athlete's running style, Dallum begins by videotaping the athlete in action wearing special markers on knees and ankles. The computer can pick up the markers from the videotape and use them to measure the athlete's actions. One of the measurements he looks for in regard to the upper leg is maximum thigh extension—how far Deboom swings his upper leg before he sets his foot down. Then he looks for maximum knee flexion—how much Deboom decreases the angle between his lower and upper leg as he swings his leg backward and forward. He also looks at the lower leg angle at foot strike.

Day 2: Cycling. Deboom mounts the stationary bicycle. While Dallum and his colleagues measure Deboom's physiological activity, biomechanist Jeff Broker videotapes the session in order to custom-build a bicycle tailored to Deboom's particular shape, size, and form. "All bodies are not the same," Broker explains. "My job is to match Tim's body to his machine. Given the tight margins of victory in Olympic events, it's critical to translate the energy from Tim to the road as effectively as possible. We look at aerodynamic drag and pedaling mechanics to find ways for Tim to reduce the drag while maintaining pedaling efficiency."

"We're always tinkering with the sport and figuring out what the interactive forces are, what the motions are. Then, with that data, we go back and we say, all right, now how do we optimize? We get very deep into the math."
JEFF BROKER
sports biomechanist

Listening to Broker describe the science of bicycle design is very much like listening to John Marshall talk about designing racing yachts. Broker measures upper and lower leg angles throughout the pedaling motion. He measures the force exerted on the pedals by each foot. And he studies the aerodynamics with the same eye for detail as an aircraft designer.

"Most of the drag is in the athlete, and the position of the athlete is paramount," explains Broker. "That's why we spend a lot of time putting the athlete in a good position, where they can still generate power. In the world championships this year and in the Olympics in Atlanta, several world records were broken, and they weren't broken because of bikes—everybody had super bikes out there. They were broken because of some innovative positions. The Italians and

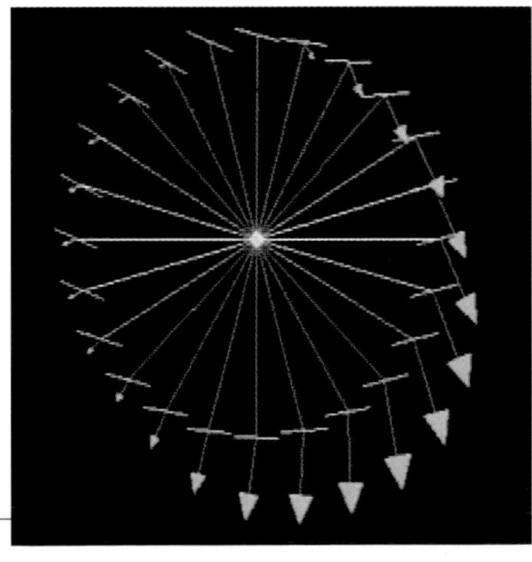

A computer analysis of Tim Deboom's pedaling motion on one pedal. The arrows represent the degree of force being applied to the pedal through the full cycle.

the Britons came up with a new position, with the arms almost straight out in front of the rider, very high up. Roughly eleven seconds were taken off a four-minute event. That's like someone running a mile eleven seconds faster than anyone else. It's unheard of. At the speed these riders are moving, aerodynamics are extremely important, and mean the difference between winning and not even placing."

Day 3: Swimming. Because continuous testing is not possible while an athlete is in the pool, Dallum relies on mathematics to monitor Deboom's performance. "Here is one of the many ways we use mathematics," Dallum explains. "We have to extrapolate what Tim's body is doing between tests. We fill in the blanks using mathematical analysis."

In order to do this, Dallum explains, "We have the athletes on a progressively faster pace for a certain distance. We measure their lactates, we measure their heart rate, we measure their perception of effort, we measure the number of strokes it takes them, we measure the frequency of their strokes. Ideally, at the peak point of the season, I can reach the optimum."

Running, cycling, swimming—once all the data are in, Dallum can start to work out Deboom's training regimen. Says Dallum: "The detailed data allows us to create a precise profile for Tim—a numerical model, if you will. With the enormous database that we have, we can compare his model to world record holders to find patterns of sameness and difference. From the comparison we can construct a unique training plan and predict how that will improve his profile over time."

Once the profile has been set up, Deboom can train by competing against his own mathematical model.

As with John Marshall and his efforts to squeeze that tiny, but oh so critical, performance advantage out of his America's Cup yacht, so too coaches like George Dallum are looking for that tiny, key edge. Referring to the science and the mathematics he uses with his triathletes, Dallum says: "What all this additional effort does is produce that last maybe five or ten percent of improvement that's necessary to take an elite athlete from fifteenth in the world championships to second or third. A lot of the work that we do in one year might produce only a one- or two-minute change in the athlete's performance, but that might be enough to move them from being a top-ten finisher to being a medalist. That's what this kind of program is about.

"Exercise is basically a stress for the body, so by applying an amount of stress that the body can accommodate, you strengthen the body. The trick is to make those increases in stress very small, incremental over a period of time. The main thing that the mathematical analysis does for us is lend more precision to that general model, to that idea that we are trying to make small incremental changes."

INSIDE THE MIND

From the aerodynamics of baseballs and golf balls to the design of tennis rackets and ocean racers, the dynamics of the triple axel, and the training of endurance athletes—these days, mathematics plays a major role in many aspects of sports.

But as any experienced coach will tell you, improvements in equipment and training on their own are not enough to guarantee success, no matter how talented the athlete. It takes the right mental attitude as well. Can mathematics be of

Expert markswoman Kim Howe wearing a voltage-measuring device and about to take a shot.

Computer analysis images of the path of the rifle end around the target area just before the marksman shoots. Right, a representative path of an expert, which is tightly focused around the bull's-eye. Opposite, a representative path of a novice, which swings wildly around the target, moving so far out of range that the path sometimes falls outside the frame of the computer image. The expert has learned to hold the rifle much steadier.

assistance in helping world-class athletes prepare mentally for competition? Brad Hatfield thinks it can.

Hatfield, a professor of psychology at the University of Maryland, is studying the brain activity of athletes to try to understand the mental processes that occur during competition. "There are a lot of quacks and nutty ideas about how to train mentally for sports," Hatfield observes. "We still don't know how to create a scientifically sound regimen for training athletes mentally. Right now, we are still learning how an athlete's brain works.

"The exercise physiologist primarily will look at the muscle, at the cardiovascular system. And most of the sports psychologists have not really addressed the brain. They primarily will ask athletes about their thoughts, or they will look at their responses on typical psychological tests, tests of personality, tests of anxiety. But there's a big, big gap between what a person reports on a psychological test and what's actually happening in their mind as they compete. We think that we can understand more about what makes an athlete perform very well, what makes for a high level of human performance, if we actually look at the brain."

Because he needs to take subtle measurements, where body motion would interfere with the measuring process, Hatfield has chosen to concentrate on the mental activity of Olympic marksmen. Though stationary during the crucial moments of competition, competitive marksmen are subject to exactly the same pressures as other athletes—the need for total concentration, complete body control, and an ability to overcome the tension that naturally accompanies top-level competition. Moreover, it is easy to measure performance very accurately—did the marksman hit the bull's-eye, or exactly how far off was the shot?

By attaching voltage-measuring devices to a marksman's head, Hatfield and his colleague Amy Haufler can measure the level of

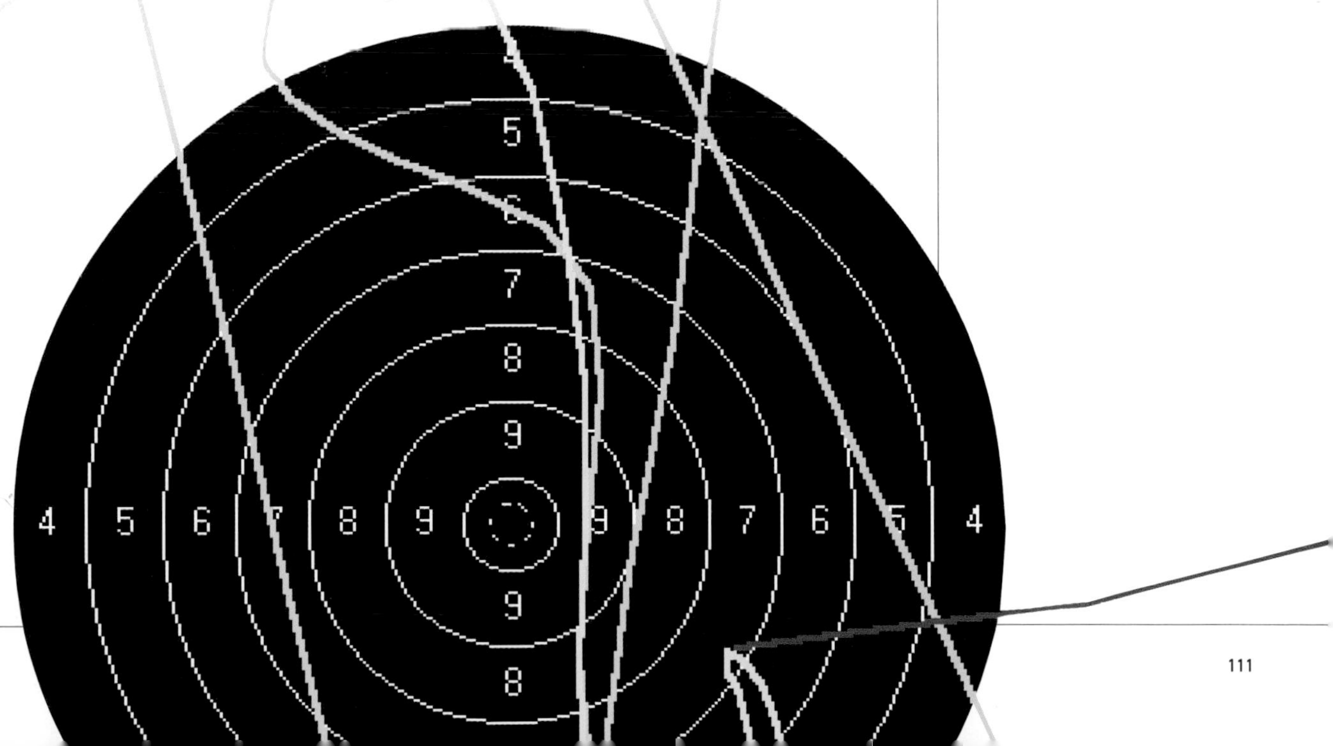

A brain map of an expert marksman, below, versus that of a novice, opposite, at one second before pulling the trigger. The view is looking down at the brain from above, with the front of the head at the top. Pink represents areas of high activity, and these maps clearly show that the expert has less left-hemisphere activity and is exerting less mental energy at the moment of shooting.

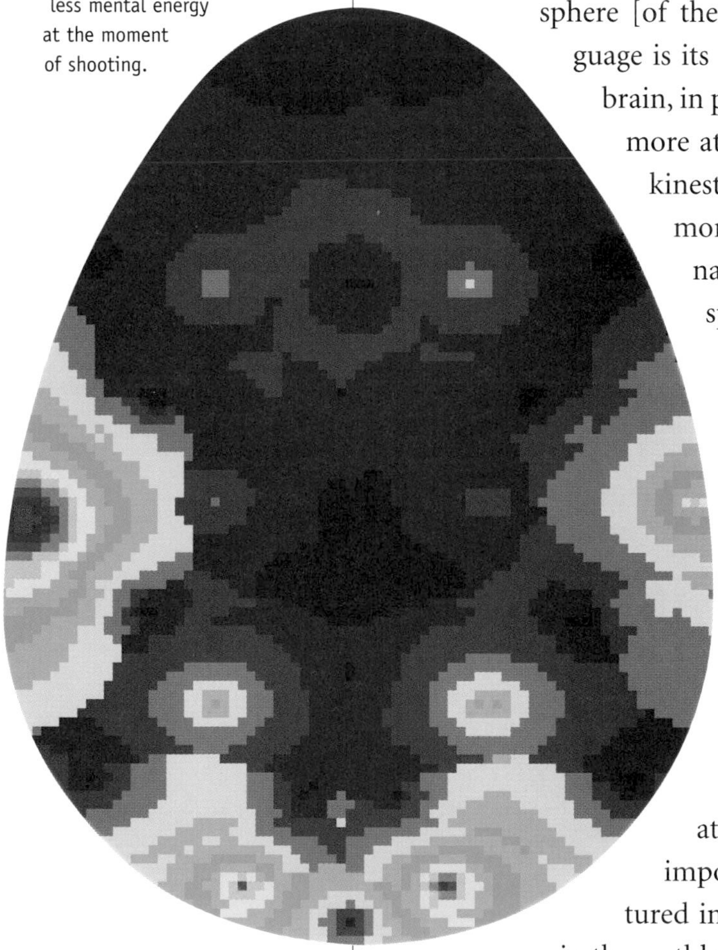

electrical activity in different parts of the brain during the various stages of competition. The measurements form what is called an electroencephalogram, or EEG, a complex array of waveforms that indicate the mental activity going on in the brain.

Hatfield explains the theory behind this approach: "We have a model of the brain that is based on the well-known principle of the difference between the two hemispheres. The left hemisphere [of the brain] is primarily for self-talk; language is its most distinguishing feature. The right brain, in popular terms, would be considered the more athletic brain. It has more sensitivity to kinesthetic or muscle awareness, it would be more involved with visual-spatial coordination where the body moves through space. With shooters we believe that the people who are at the lower end of the skill level will show increased activation on the left hemisphere, where they are thinking too much about what they do. As we look at people who are very highly skilled, our model would predict much less activation in the left hemisphere. That shift to the right hemisphere dominance is what we believe underlies focused concentration. Of course, athletes talk about focus as of supreme importance, and we believe it can be captured in some of the activity that we measure in these athletes' brains."

To interpret the EEG readings, Hatfield and Haufler use a mathematical technique known as Fourier analysis, which takes a complex waveform and breaks it up into a collection of much simpler waveforms, called sinusoidal waveforms. The process is analogous to taking, say, a cake and breaking it up into its basic ingredients—flour, eggs, milk, butter, salt, and so forth. Of course, we know that if we start with a baked cake, we are unable to break it down into the original ingredients (though we might be able to analyze it chemically to find out what those ingredients were). In the case of waveforms, however, in the nineteenth century, the French mathematician Jacques Fourier showed how to break a complex waveform into its basic wave ingredients. These days, Fourier's method, known as Fourier analysis, is widely used in science, engineering, communications, and several other walks of life. (Used "backward," to build up a complex waveform from simple constituent waves, Fourier analysis is the basis of electronic music synthesizers.)

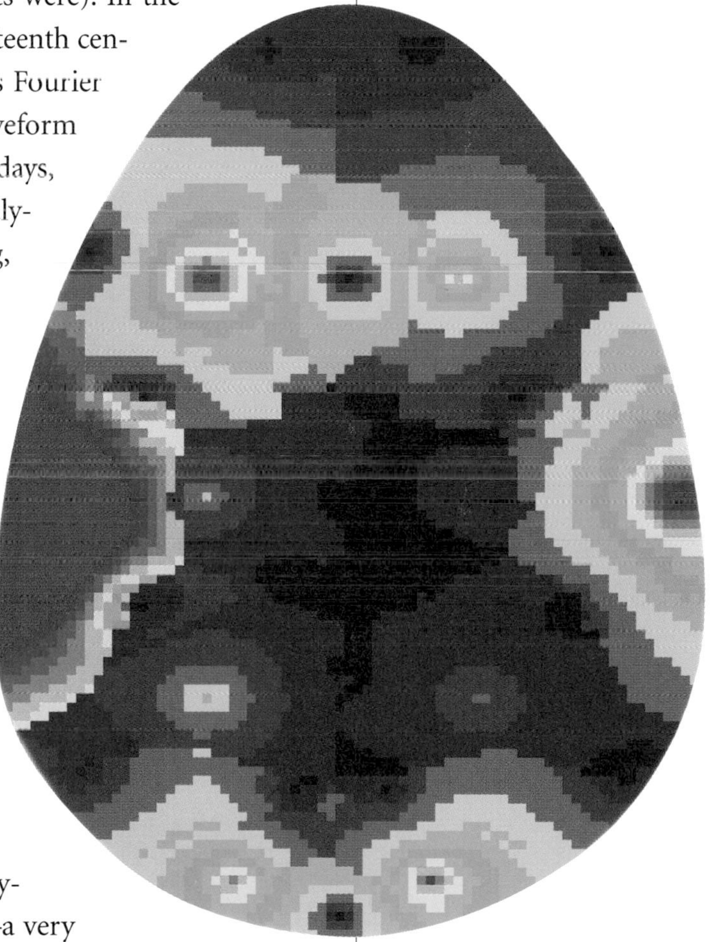

For Hatfield, Fourier analysis is typical of the way mathematics works: "That's the beauty of mathematics," he says. "It takes the universe and breaks it down into some simple principles, or it takes nature and it breaks it down into some simple principles. That's exactly what we're trying to do in our work with the mind—a very complex organ."

Applied to the EEG readings from an athlete's brain, Fourier

"Some of the greatest developments in sports psychology will come from developments in computer science and electrical engineering, because they will allow the mental processes to become measurable— much more visible, much more real."
BRAD HATFIELD
sports psychologist

analysis gives a profile of the different activities going on. For example, what is called alpha wave activity—8 to 13 hertz—is characterized as a very relaxed but alert state of mind. Beta wave activity—14 to 33 hertz—is characteristic of very active cognitive processing.

By measuring the brain activity in marksmen and analyzing the results using Fourier analysis, Hatfield and Haufler can start to map the different areas of the brain that are active during the various stages of the competition, and the kind of activity that is going on in those areas. By then comparing these maps of brain activity in expert shooters to brain maps of novice shooters they have made a fascinating discovery. The brain of a novice shooter shows a good deal of higher-energy beta wave activity, especially in the left side of the brain, which governs analytical thinking. The novice is doing quite a bit of higher-energy cognitive work in the course of making a shot. The brain map of an expert marksman, however, shows much less higher-energy activity, and the difference from the novice is especially large in the left side of the brain.

The expert responds to the challenge of shooting with much less analytical processing, and makes a shot with much greater mental efficiency. As Hatfield explains, "People call it going with the flow; alpha states; just doing it. That's the anecdotal description. We're trying to quantify what's actually going on."

Hatfield admits that this research is still very much in its infancy, the goal of being able to use these results to improve performance perhaps decades away. The goal itself is, of course, very reminiscent of the way coach George Dallum uses physical and physiological data to develop training programs for endurance athletes. "When we can model an athlete's brain activity accurately," says Hatfield, "then he or she can train against his or her model. We may be fifty years away from having mental training be as concrete as it is in the physical realm. But with the development of the mathematical model and the evolution of technology comes the possibility of an athlete learning to control and optimize brain activity the way we now train to improve metabolic and mechanical performance."

Through the work of psychologists like Hatfield, mathematics, to many the supreme product of the human mind, is at last starting to help us understand the very mind that produced it. And in so doing, Hatfield may help athletes to improve their performance, to reach heights never attained before.

Some people see the modern scientific approach to athletics as somehow "impure." They look at all the charts and tables, at the scientific equipment, at the physiological measurements and the pages of computer printouts, and they feel that the original goal of athletics has been lost. But is today's mathematically based approach really any different from the athletics of old? Have we lost the original ideal? Olympic triathlete coach George Dallum doesn't think so: "An elite athlete is an ideal expression of human potential. In ancient Greece they were revered as gods. Milon of Croton was an early Olympian who trained by carrying a newborn calf a certain distance every day. As the calf grew into a fully grown cow, Milon became stronger. That's the concept of what we are trying to do. We are just being more mathematically precise."

Chapter 5

THE SHAPE OF THE WORLD

In years when the rains were particularly hard, the ancient Egyptians living in the Nile Valley faced a recurrent problem. The river would burst its banks and flood the surrounding plains, washing away the markers that the farmers used to indicate the boundaries of their land. To overcome this problem, they needed a way to accurately reconstruct the obliterated boundary lines. That meant taking measurements of the earth's surface and recording them in the form of a map. Thus was geometry born—or so the story goes. In any event the word *geometry* comes from the two Greek words *geo,* meaning earth, and *metros,* meaning measurement.

From its beginnings five thousand years ago in ancient Egypt, geometry developed into a rich and powerful branch of mathematics, having many applications.

From those same beginnings, modern mapmaking—the official term is *cartography*—developed, as people sought ever more reliable ways to answer the questions, Where am I? and How do I get where I want to go?

Willem Janzoon Blaeu's world map of 1635 with border panels depicting the four elements, the seven known planets, the four seasons, and the seven wonders of the world.

Those questions came down to knowing the shape of the earth—the shapes of the landmasses, the shapes of the different regions and countries, the positions and shapes of the rivers, where the mountain ranges are, the layout of the roads, and so on.

Today, mapmakers are still making use of geometry to draw maps. Nigel Holmes is a present-day mapmaker. His maps appear every week in newspapers and magazines such as *The New York Times, Time,* and *Sports Illustrated.* The maps he draws show readers where the explosion happened, where the plane crashed, where the armed conflict is.

In drawing his maps, Holmes faces the same geometry problem that all mapmakers have faced throughout the ages. It's a two-part problem, often made more difficult because one part can affect the other. The first part is scale. A mapmaker may have to represent anything from a city block to an entire country, or even the whole world, on a map measuring a few inches across. To do this, the map has to be drawn to scale, with, say, one inch on the map corresponding to a quarter of a mile of city streets or a thousand miles of the earth's surface.

In the case of a map of a city, or even a small country, drawing a map "to scale" is not difficult. You just multiply all measurements taken on the ground by a fixed "scale factor." The hard question is to decide what details to include in the map and what to leave out. That's not a mathematical question; it depends on the purpose to which the map will be put.

It's when the map covers a significant portion of the earth's surface that things get tricky. That's when you face the second part of the mapmaker's problem: the earth's curvature. How do you represent features of a curved surface on the flat surface of a map?

Holmes likes to explain the problem by using an orange to represent the earth. To obtain a map of the earth's surface, the most obvious approach is to remove the peel and lay it flat. But no matter how hard you try, the orange peel stays curved. It is in order to overcome the orange-peel problem that mapmakers turn to mathematics.

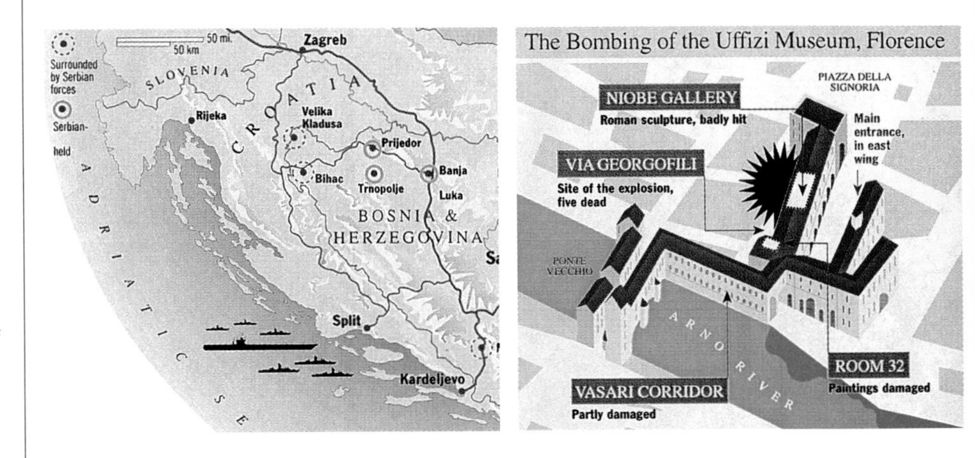

These two maps by Nigel Holmes pose different challenges of scale. On the near right, Holmes must cover a sizable territory to depict the strategic problems in attempting to secure the city of Sarajevo during the Bosnian-Serbian war. On the far right, he must focus on one building, the Uffizi Museum in Florence, Italy, in order to depict the section of the museum damaged by a bomb.

The challenge facing the cartographer, then, is to represent the curved surface of the earth, or a portion of it, on a flat sheet of paper. Mathematicians call such a representation a "projection"—they say the earth's surface is projected onto a flat plane. Ideally, a cartographer wants to represent accurately those geographic features that are important in a map—direction, shape, scaled distance, and scaled surface area. In other words, to be really useful, a map should tell you which direction to go to get from *A* to *B*, what is the shape of a particular region *C* (state, country, topographic region, etc.), how far it is from *A* to *B*, and how big is region *C*. However, because of the orange peel problem, it is not possible to find a projection that preserves all four of these features at once. The cartographer has to make a choice, depending on the purpose the map is intended to serve. Once that choice has been made, the mapmaker then turns to geometry to find the appropriate projection—a formula that translates points from one geometric shape to another, from a sphere to a flat plane.

TO GO FROM *A* TO *B* OR TO FEED THE WORLD?

A familiar example of a projection, found in many school atlases in Europe and North America, is the Mercator projection, introduced by the Flemish geographer Gerardus Mercator in 1569 in order to provide sailors with a reliable navigational tool.

Back in the sixteenth century, long before the days of modern electronic navigational aids, if you were a sailor, the most important feature in a map was compass direction. It was even more important than distance. Accordingly, Mercator drew his map so that it was accurate in terms of direction: if a sailor drew a straight line on the map from, say, Gibraltar in the Mediterranean to Boston in the New World, then the line represented the path from Gibraltar to Boston that followed a constant compass bearing. What is more, at any point on the line, indeed at any point on Mercator's map, north is directly above the point and south directly below.

How did Mercator draw such a map? The key to drawing any map is the grid system: how you represent on the map the lines of latitude and longitude that curve around the globe. Once you have drawn the grid system, all that remains is to transfer the details within each grid region of the globe to the corresponding grid region of the map. The grid system determines both the scale and the degree of distortion that will arise because of the orange-peel problem.

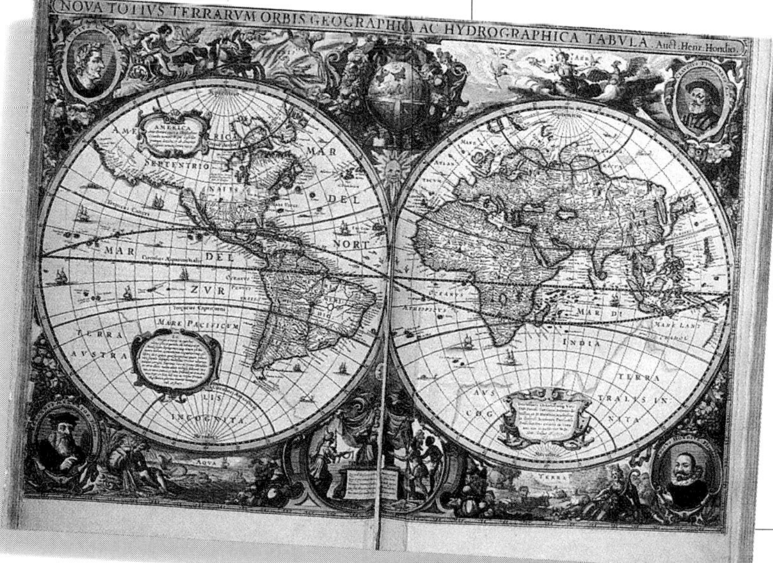

Mercator's famous map of the world, published in 1569. Note the distortions in the relative sizes and shapes of the continents.

Mercator's map uses a rectangular grid: lines of longitude are drawn vertically; lines of

Right, a Mercator projection map, opposite (top), an oval projection—this one a Robinson projection—and opposite (bottom), a Peters projection. Note especially that in the Mercator map, Greenland appears larger than Africa, whereas in the Peters projection, Greenland appears in its true size. In the Peters projection, however, the continents appear in misshapen forms, and so the Robinson projection in the middle is more visually pleasing. With every style of projection, there are positive and negative factors the cartographer must weigh against one another.

latitude are drawn horizontally. For town plans and maps that represent at most a few hundred miles, rectangular grid maps are perfectly adequate in all respects. But for a map of the world, they present a significant problem: they distort shape. The further you get from the equator, the greater the distortion, as small regions near the poles get stretched out farther and farther across the width of the map. For example, on the Mercator map, Greenland appears to be a huge continent, very wide at the top, but in reality it has the same surface area as Mexico, which looks fairly tiny on the map.

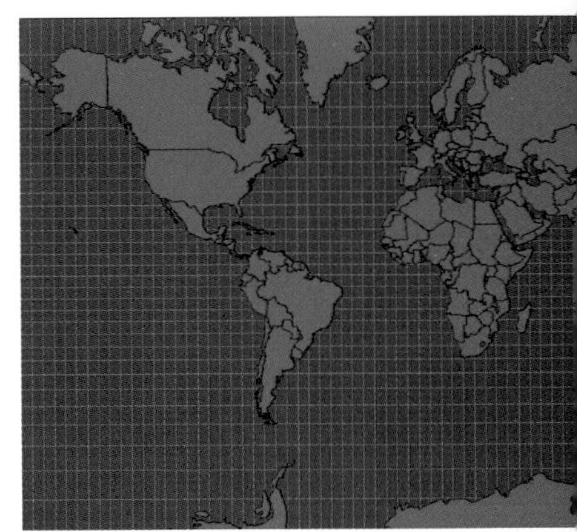

To obtain a rectangular map that was true to compass bearings, and thus could be used by navigators, Mercator compensated for the increasing horizontal stretching that occurs the farther you get from the equator by steadily increasing the vertical distance between the lines of latitude. "I have gradually extended the degrees of latitude in the directions of both Poles in the same proportion in which the parallels of latitude increase in their relation to the equator," he wrote. Though it would have been possible to draw his grid system by calculating the different distances between the lines of latitude, Mercator did it geometrically, using techniques of technical drawing.

While useful for sailors in the days when navigation was done largely by compass, the Mercator map has a number of disadvantages. For instance, in today's world, perhaps the most common reason people consult a map of the world is not navigational but social—to gain an overall sense of the world and to locate world events. For those kinds of uses, compass direction is irrelevant, but shape and area are arguably very important.

For example, the Mercator map shows Greenland about the same size as the African continent. But in reality, Africa, with an area of 30 million square kilometers, is nearly fifteen times larger than Greenland, which has an area of 2.1 million square kilometers. The Mercator map also shows Africa as slightly smaller than North America, giving the impression that the two continents are of comparable size. But again, the reality is quite different. The area of North America is 19 million square kilometers. Africa is half as big again! In fact, a glance at a globe will show that Africa dwarfs practically every other landmass!

Thus, when it comes to trying to get a general sense of the size of the world, the Mercator map is extremely misleading. In particular, the distortions that result from the mathematics Mercator used in order to get the right compass bearing tend to obscure the huge problems that can arise in feeding the peoples of the African continent.

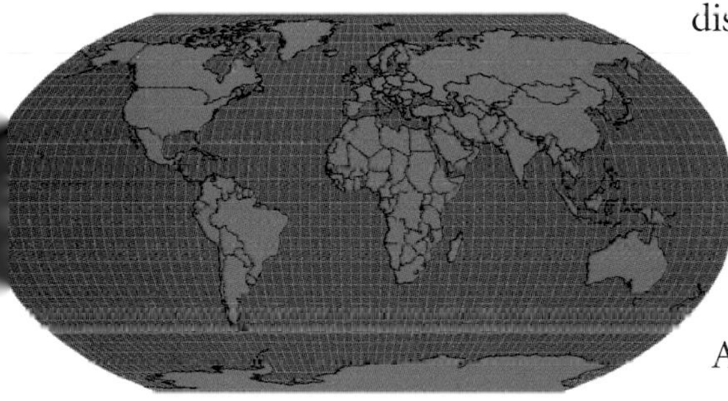

This was precisely the point made by the German cartographer Arno Peters, who, in 1983, produced a rectangular-grid map of the world that is faithful to land area. These days, argued Peters, the most important feature to preserve in a global map is land area. Next comes shape. Forget compass bearings altogether. The Peters map distorts shape and distance (and compass direction), but is accurate in terms of surface area. To obtain his map, Peters had to resort to some fairly intricate mathematics—there is no simple geometric projection that will produce a map representing area accurately.

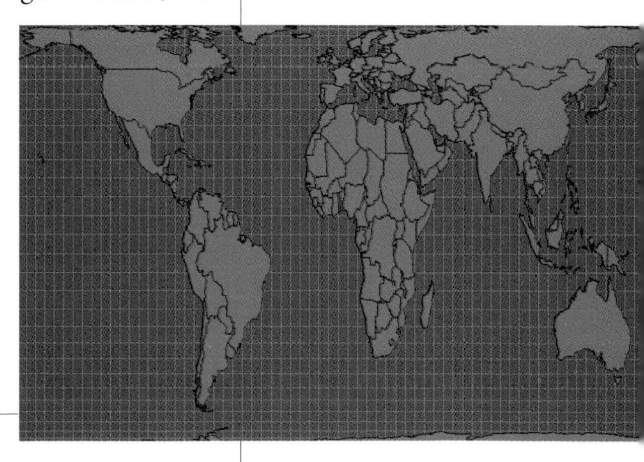

The introduction of the Peters map highlighted a feature of mathematics that professionals are well aware of but that outsiders sometimes overlook. Mathematics is completely accurate and precise. It does for us exactly what we tell it to do. But if we forget the assumptions we make before we start to use the mathematics, if we forget what features we decided to ignore, then the result can be misleading.

Because of the orange-peel problem, in order to represent the round earth on a flat surface, the cartographer has to decide what features to preserve, and then use the appropriate mathematical transformation. The resulting map gives a view of the world shaped by that mathematical transformation, and that view can affect the way people approach the world's problems. Mercator's map was highly reliable for the ocean navigator. But it is woefully inadequate for giving any sense of how big the different continents and countries are.

THE VIEW FROM ABOVE

These days, a lot of the mathematics involved in making maps is hidden from the user, embedded in technology, as cartographers use satellite data and computers to draw ever more accurate maps. For instance, by taking aerial photographs and elevation measurements from equipment in orbiting satellites, the United States Defense Department has developed virtual-reality flight-training simulators for almost every area of dry land in the world. Using these highly accurate—and very lifelike—representations of the terrain, military pilots can plan operations and train for missions before they ever step into a real aircraft.

Vic Kuchar of the United States Defense Mapping Agency describes this new kind of mapmaking this way: "The terrain data is the mathematical model that gives us the hills and the valleys—a mathematical representation of the earth which allows us to do several things, such as locating ourselves precisely on the earth. Also, we overlay imagery and tie it down precisely, put in our intelligence information, our operational information, and our planning information, and get it to the people who have to have it." He adds: "The mathematics is transparent to the user, but it's the basis for everything that we do."

Satellite data is also what lies behind the "global positioning system," or GPS. The GPS is a recent innovation that can tell you your exact position on the earth's surface—down to the street you are on, in some cases.

The proud father of the GPS, the man who directed the program that brought it to life, is Brad Parkinson. As Parkinson explains, GPS uses a new way to determine location, quite different from the method used by generations of navigators since ancient times: "To determine their position, the ancient navigators would measure the

A still from the terrain simulation system used by the Defense Mapping Agency to assist in the negotiation of the Bosnian peace agreement in Dayton, Ohio. This system let negotiators view the disputed territory in great detail from their remote location in Dayton and draw and redraw boundary lines. This view is as seen from forty-eight thousand feet, and the green line delineates the Gorazde corridor negotiated to maintain access to Sarajevo.

elevations of stars. To give a particular angle for a specific star, you must be on a certain circle on the earth, which you can work out mathematically. By looking at several stars, they found the intersection of all those circles, and that gave them their location. GPS doesn't do that at all. The mathematics is quite different and the results turn out to be quite a bit better."

For instance, in Atlanta, Georgia, a GPS navigation system currently under testing links a car's real-time location to a digital map stored in a computer on the car. On a computer screen mounted on the dashboard, the driver can see the car's location at a glance. As the car makes its way around the city, an onboard receiver detects radio signals from satellites eleven thousand miles overhead. A clock in each satellite is synchronized with all the others, and a coded time signal is broadcast to the earth below. The receiver compares these signals to its own clock to measure the distance to each satellite. The distance defines an imaginary sphere around the satellite, and the intersection of all the spheres marks the car's latitude, longitude, and altitude on the earth's surface.

To locate the position of the car, the location of the satellites must be accurately pinpointed at every moment. At Falcon Air Force Base in Colorado, satellites are monitored around the clock to provide this information. Each satellite carries the information giving its location at any moment: latitude, longitude, and altitude. At the heart of this positioning system is a mathematical observation made three hundred years ago by the astronomer Johannes Kepler. Kepler examined the way planetary objects move in orbit. One of a number of observations he made was that the orbit of a planet will be an ellipse.

Kepler's observation is not true just for planets; it also holds for a twentieth-century satellite—almost. The complicating factor for an orbiting satellite is that, because of its small size, it drifts up and down several miles as it travels through its orbit. Over tall mountains, the pull of gravity increases, dragging the satellite down; over oceans, the pull is less, and the satellite drifts upward. The amount of vertical drift has to be calculated. This can be done so precisely that a GPS satellite can calculate its own latitude, longitude, and altitude at any moment to an accuracy of a few meters.

A Global Positioning System monitor mounted on the dashboard of a car. The system enables drivers to determine their precise location instantly.

As Parkinson explains, the satellite broadcasts its position down to the car below: "In essence, the satellite is hollering down, 'I am here, I am here,' and the user says, 'Yes, I know you're there and I know what time it is, I can sort out where I am.'"

With GPS, timing is crucial. The onboard clocks are so accurate that they lose no more than one second in three thousand years. The GPS needs this kind of accuracy because it is timing a signal that is moving at the speed of light, 186,000 miles per second. That means that one billionth of a second is equivalent to one foot.

Each satellite in the Global Positioning System covers a certain territory of the earth's surface and reports its data to a radio monitor, represented here by the large satellite dish, around the clock.

For Parkinson, the GPS's designer, the system represents yet another remarkable example of the usefulness of mathematics. "Math is at the heart of this system," he observes. Yet he is aware that most users of the system will never see the mathematics—they will simply take the system for granted. The all-important mathematics will remain invisible.

As Parkinson observes, "Modern man takes a lot of things for granted. Man in the seventeenth century could no more conceive of going faster than about seven miles an hour on a horse than flying to the moon. What has happened now is we have a new power—the power is knowing where you are. The mathematics that's embedded in this, as elegant and wonderful as it is, is another example of something that modern man will take for granted. It's invisible. Someone can push a button. The math is embedded in a little chip. It gives us our answer: here we are."

THE MOUNTAIN MAN

Brad Washburn had neither satellite data nor computers when he set out to survey the Yukon Territory in 1937. To obtain aerial data, he went up in a small airplane, tied a rope around his waist, leaned out of the open door—the air temperature was around twenty-five degrees below freezing—pointed a camera at the ground below, and took photographs. Using the photographs to identify key survey points, the rest of the survey work was all done on foot.

It's hard to realize that as recently as the late 1930s, large parts of North America had never been surveyed—there simply were no maps. One such unmapped area was a five-thousand-square-mile region of the Yukon, near the Canadian border. Maps at the time simply left the region white, with the legend "High mountains, 10 to 12 thousand feet high." That's all that was known about the area. Seeing the opportunity for an adventure, Washburn, a keen explorer, approached *National Geographic* and asked them if they were prepared to finance an expedition to survey the region. They were.

Washburn started with the aerial photography. It was the first time anyone had used aerial photography to draw maps, and Washburn's first experience with flying. By drawing a network of angles on the photographs, Washburn and his team were able to determine the approximate locations of the major mountains and their rough heights. Then they assembled all the equipment they would need and set off on foot. The expedition lasted three months, from March to June.

The key technique they used to determine distances and heights was called triangulation. This involved taking two "base points" a known distance apart, and measuring angles from those base points to nearby peaks. The unknown distances and heights—all sides of right triangles—could then be computed by fairly straightforward calculations using the methods of trigonometry.

The aerial photographs provided Washburn and his team with two advantages not available to earlier mapmakers. First, by starting with the approximate data obtained from the photographs, they were able to proceed much more quickly on the ground than would otherwise have been the case. Second, once they had the measurements from the ground survey, they were able to compare them with the photographic data in order to draw the final maps.

By selecting one base point directly vertical to the point at the top of the mountain, a right triangle can be drawn connecting the three points. The surveyor knows the distance between the two base points, which is easily measured because the two points are on level ground. The geometry of right triangles allows the surveyor to then use simple calculations to come up with the height of the mountain peak.

The Yukon expedition was part of a major advance in the way maps are made. Together with his wife, Barbara, Washburn went on to use the new aerial photographic technique to map a number of the world's highest peaks. Their work helped start a rapid evolution toward more and more sophisticated mapmaking methods.

Looking back on his life as an explorer and mapmaker, Washburn comments: "I think the fundamental difference between today and a hundred years ago—or a thousand years ago—is that the explorer today has to be a much more technically trained person than you had to be in the old days. Back then, you could just snowshoe into areas that nobody had ever seen before. These days we are living in a technological world."

That technological world is built on a foundation of mathematics, a foundation that for the most part remains hidden from view, an invisible universe that can be seen and explored only by donning a pair of mathematical spectacles.

BENEATH THE OCEANS

Dawn Wright also explores a universe that is normally hidden from view: the ocean floor. She is a fairly new type of scientist: an oceanographer.

What led to her initial interest in the ocean? Wright explains: "I grew up in Maui, Hawaii, so I was always surrounded by the ocean. I did a lot of surfing and snorkeling, and I just loved it. So I thought, well, this is great, I wish I could do it full time. Then, when I discovered that oceanography was a science and a science I could study and become proficient in, I thought, well, why not. From around age eight, I decided I was going to be an oceanographer. So, I got a bachelor's degree in geology, a master's degree in oceanography, and a Ph.D. degree with an oceanography emphasis. That's normally the way you have to go."

Like other scientists, Wright uses mathematics to see what would otherwise remain invisible. "The areas that I study are anywhere

Ocean scientist Dawn Wright, front and center, with her colleagues aboard a research vessel.

from one and a half to three miles in depth," she explains. "It's not like I can just go out of my front door and go to the Juan de Fuca Ridge and walk around for a little while and gather some data and go back to my lab." The Juan de Fuca Ridge is a spectacular mountain range that lies beneath the surface of the North Pacific Ocean.

Wright explains her work: "I'm interested in very active areas where there are lots of volcanic eruptions, lots of earthquakes. There's a tremendous amount of heat released from the surface of the earth. It's very important in terms of global climate. It's important for me to be able to take back maps of what these areas look like. Because I don't know when I'll be able to get back again."

"I think people realize that oceans are important. I think that what's missing, though, are the details of why the oceans are important."

DAWN WRIGHT
oceanographer

Wright uses sound to chart the ocean floor. An instrument mounted on the bottom of the research ship sends out pulses of sound, which bounce back up to the surface to be picked up again by the same device. Since the speed of sound in water is known, by measuring the length of time it takes for a sound pulse to travel down to the ocean bed and back again, the depth of the ocean floor can be calculated. By making such measurements as the ship travels up and down on the surface, a chart of the changing depths can be drawn, showing the terrain. The survey vessel uses Brad Parkinson's global positioning system (GPS) to determine its exact location at each moment. Mathematics is also used to "fill in" details that are missed by the echo system—a technique known as interpolation. The entire process is automated, the mathematics hidden from the user.

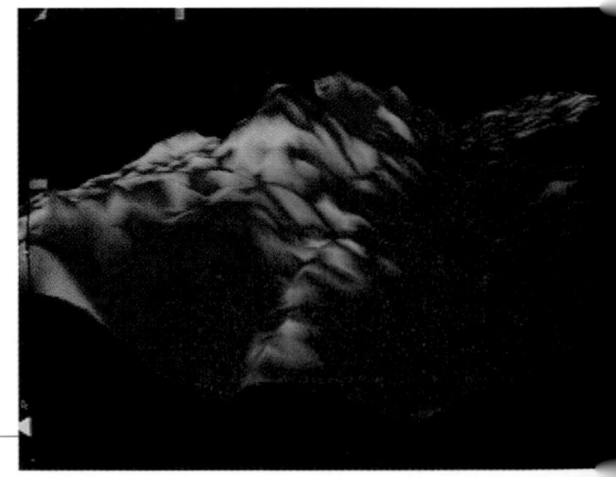

For Wright, the data she collects

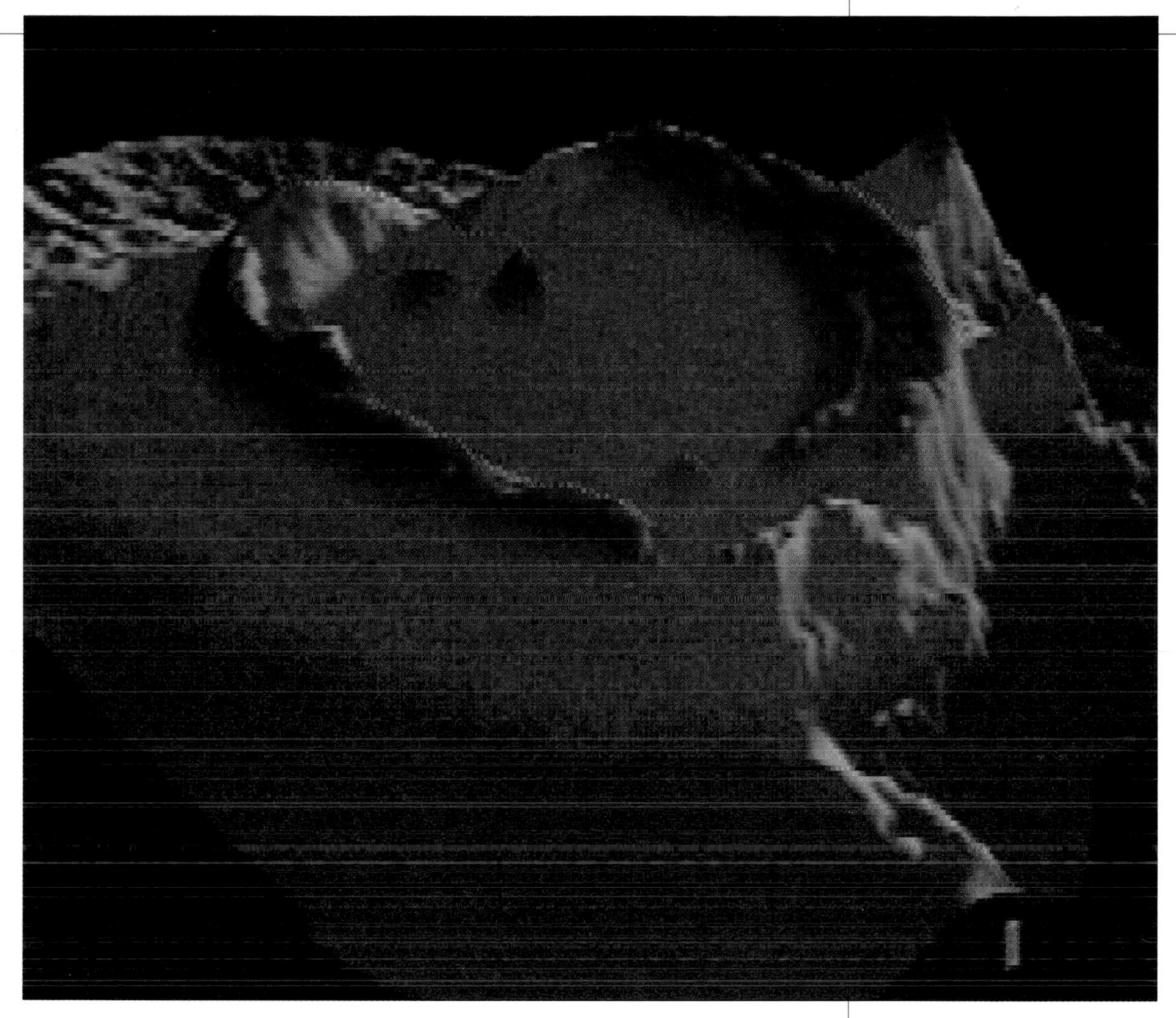

Computer-generated images of a volcano (above) and of a massive canyon (opposite bottom) on the ocean floor.

can be fascinating. Like Brad Washburn, who charted the unknown Yukon Territory, Wright is an explorer who ventures into new terrain. "There are incredible trenches and canyons like the Grand Canyon," Wright says, "there are these incredible mountains. Every time I look at a map I see something a little different. It's very exotic. I know I am making maps of places people have never been to and probably will never be able to go to. It's basic exploration of uncharted territory."

It is a world that Wright can see only with the help of mathematics.

IS THE UNIVERSE BENT?

Dawn Wright needs mathematics in order to see the seabed. Astronomer Robert Kirshner uses mathematics to see the far reaches of the universe. One of the fundamental questions he wants to answer is this: Is the universe flat or curved?

Two thousand years ago, the ancient Greek mathematicians asked the same question about the earth. By means of mathematics, they were able to deduce that the earth must be roughly spherical in shape, long before twentieth-century spacecraft would be able to send us pictures of our planet taken from outer space. In fact, in 228 B.C., the Greek mathematician Eratosthenes calculated the diameter of the earth with amazing accuracy (within 1 percent), using just two simple observations of the elevation of the sun in the sky and some elementary trigonometry.

Today, we take it for granted that the earth is spherical. But as geometer Robert Osserman remarks, such acceptance did not come easily to mankind. "For us it's such a commonplace, it's hard to realize what a leap it was. To picture the earth as a gigantic ball floating in space, with people on the opposite side hanging upside down by their heels—that was very, very hard to conceptualize." Like Kirshner, Osserman—who works at the Mathematical Sciences Research Institute in Berkeley, California—is interested in knowing whether the universe is flat or curved.

Astronomer Robert Kirshner.

The idea of a curved space started with the nineteenth-century mathematician Georg Friedrich Riemann. In effect, what Riemann said to himself was this: The surface of the earth is a two-dimensional object that curves around and comes back on itself. If you start on a journey from anywhere on the earth and keep traveling (on the surface) in exactly the same direction, then eventually you will circumvent the globe and come back to your starting point. Perhaps the same is true of three-dimensional space. Maybe if you set off on a journey through space and keep going in the same direction, you will eventually come back to your starting point. This would mean that space itself is "curved," though the idea of a curved space is hard to visualize.

In this image, curved space—represented by the cone in the blue grid is being created by an unseen body with a very strong gravitational pull, perhaps a black hole, that is located at the point of the cone and is warping space and pulling the colored, planetlike globes toward itself.

On the other hand, if you find it hard to grasp the idea of a curved space coming back on itself, then the alternative is just as mind-boggling. If space does not curve around and close up on itself, then the universe must be infinite.

Kirshner begins his investigations into the curvature of space with the masses of astronomical data obtained using large, powerful telescopes capable of peering into the far reaches of the known universe. Just as you need to be able to see a significantly large portion of the earth's surface to perceive its curvature, so too you need to see a sufficiently large portion of space to detect any curvature it might have.

A spacecraft eight hundred miles above the earth can take a pretty good photograph that shows the curved shape of the earth's surface. Kirshner needs to look at a portion of the universe 5 billion light-years away in order to stand any chance of detecting any curvature. Since the earth is only about 4.5 billion years old, that means the light signals Kirshner examines started on their journey toward the earth long before the earth even existed. In other words, he needs to look into the distant past. Only mathematics, coupled with powerful telescopes, has that kind of reach.

Kirshner explains the key idea behind his approach: "You can take a picture of a star or galaxy and you can see its shape, how it's distributed in the sky. But one of the more powerful tools we have is to take that light and run it through a prism or a grating that spreads the light into a little rainbow, and then measure how much light there is of each color—how much blue light, how much green light, how much red light, and on into the infrared."

The discovery that ordinary white light is actually a mixture of light of different colors, into which it is broken up when it passes through a prism, was made by the great British mathematician of the seventeenth century, Sir Isaac Newton. The explanation of how the prism breaks the light into its component colors is relatively simple.

White light passing through a prism splits into its component spectral rays, producing the color spectrum that is visible to the human eye.

Light travels in waves. The distance between successive crests in any kind of wave is called the wavelength. Different wavelengths of light correspond to different colors. When a light wave passes through a prism, it is bent. The amount by which it is bent depends on the wavelength; the shorter the wavelength, the more the light is bent. As a result, when white light passes through a prism, the differently colored component waves are bent different amounts. Thus, what emerges from the prism is not a single white light but a spectrum of different colors, arranged in bands, from the lower wavelength red, through yellow and blue, and up to the higher wavelength violet.

Rainbows are formed in the same way. Tiny particles of water left suspended in the air after a rain shower act as miniature prisms that split the sunlight into its component colors.

What makes Newton's discovery useful to Kirshner is an observation made by the German physicist Christian Johann Doppler in 1842, known nowadays as the Doppler effect. Doppler observed that as the source of a sound passes you—say, a car traveling toward you along a road, passing you by, and then receding into the distance—the pitch of the sound it makes seems to drop as the source passes you and starts to move away. The effect is especially noticeable when a police car or ambulance passes with its siren or bell sounding. The pitch of the warning sound is higher when the vehicle is approaching you than when the vehicle is receding away from you.

To demonstrate his observation, Doppler, who was evidently something of a showman, arranged for a chorus of trumpeters to be whisked past a crowd of spectators in an open railroad car.

The Doppler effect is easily explained in terms of wavelength. The shorter the wavelength of a sound wave, the higher the pitch of the sound as detected by our ears. When a vehicle approaches you, in between successive crests of the sound wave from its engine or siren, the vehicle will draw closer to you, and thus the next wave crest has a shorter distance to travel in order to reach you. The net effect is that the distance between the successive crests is shorter, and the wavelength of the sound wave that reaches your ears is shortened. The pitch rises above the level it would have been if the source had been stationary. When the vehicle has passed you and is receding away from you, the opposite effect is produced. The motion away from you effectively lengthens the wavelength of the sound wave reaching your ears, and you detect a lower pitch.

"The great book of nature can be read only by those who know the language in which it was written. And this language is mathematics."

GALILEO GALILEI
seventeenth-century astronomer

For a given sound source moving directly toward you or away from you, say, an ambulance with its siren going, a mathematical calculation tells you exactly how the pitch (that is, the wavelength) changes with the speed of the source. Thus, knowing the speed of sound, the Doppler effect can be used to calculate the speed at which the source is receding from you.

The Doppler effect also occurs with light. Light from a source moving toward us has its wavelength shortened, and the light appears more blue—a phenomenon known as a "blue shift." Light from a source moving away from us appears redder (has a longer wavelength)—the "red shift."

Since the start of the twentieth century, scientists have known that the universe is expanding. All the stars and galaxies we see in the sky are rushing away from us. This is thought to be a result of the universe beginning with the Big Bang many billions of years ago, a cosmic explosion that simultaneously created matter and sent it hurtling outward, a process that is still going on. The farther away from us a particular star or galaxy is, the faster its rate of recession from us. As a result, the farther out into the universe we look, the more the light from the stars we see is shifted toward the red end of the spectrum. Just as with sound waves, by measuring the red shift of a particular distant star or galaxy, astronomers can calculate the speed at which the star or galaxy is moving away from us, and can thus calculate the present distance between us and the object.

Using this red shift effect, Kirshner and his colleagues have calculated the distance to about twenty-five thousand galaxies observed systematically in thin slices of the night sky. They have been able to draw a map of the universe, with distances marked.

On the basis of the map, Kirshner hopes to be able to measure the curvature of space, and see whether it is nonzero (that is, see if

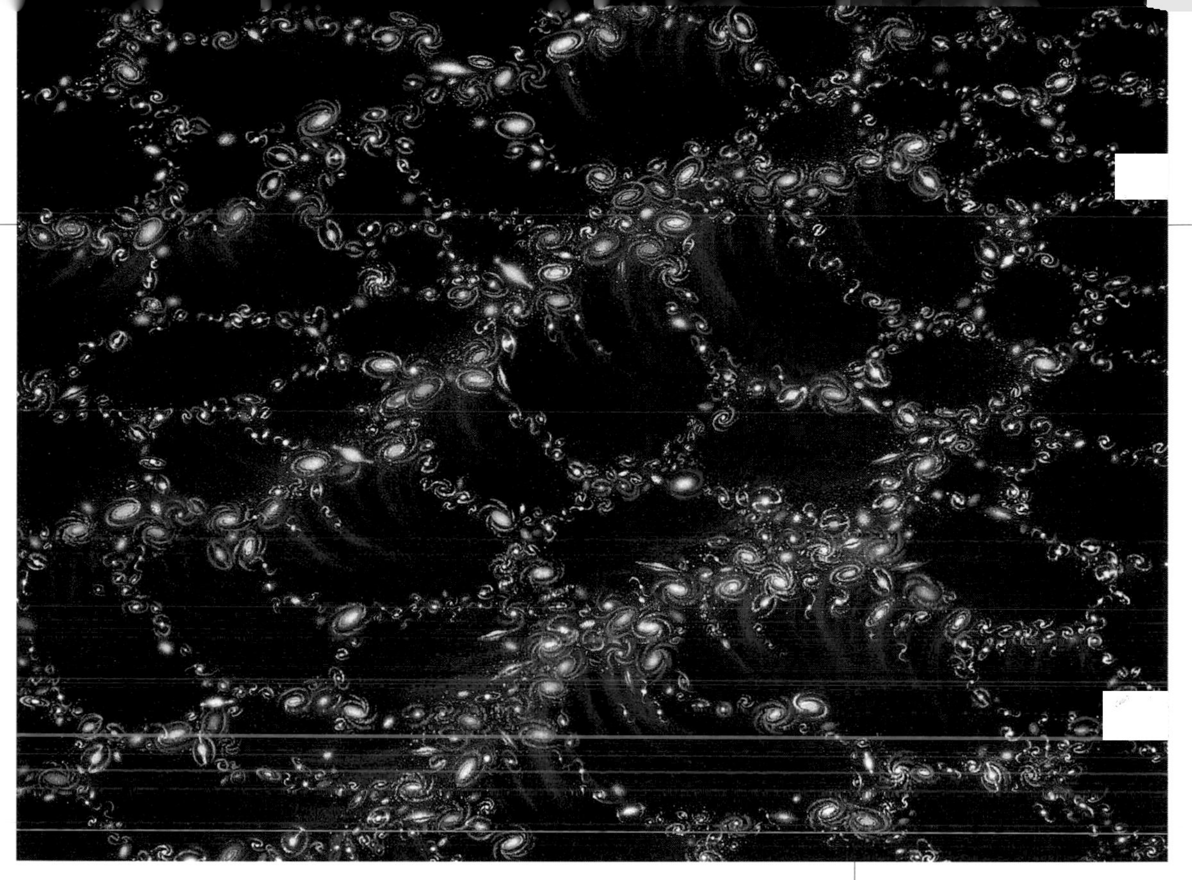

space really is curved). To do that, he focuses attention on the so-called supernovae. A supernova is a star that has burned up almost all its nuclear fuel and is reaching the end of its life, but which, instead of simply dying away, decides to "go out with a bang" lit erally. It uses all its remaining fuel to explode in a gigantic nuclear explosion, lighting up the heavens with a fireball as bright as a billion suns. The explosion typically lasts about a month.

In this computer-simulated image of one possible distribution of the galaxies, we see that galaxies are not evenly distributed throughout the universe but are collected into clusters, which in turn form superclusters and are connected in a weblike structure. This clustering leaves voids between clusters that can be hundreds of millions of light-years apart. The uneven distribution of the galaxies is a result of the Big Bang that scientists believe created the universe.

Using the Doppler effect, by examining the color of the light from a supernova that reaches his telescope, Kirshner can calculate the distance of the supernova from the earth.

By repeating the same process for many different supernovae, he can plot the relationship between distance and the brightness of the supernovae. In a flat space, the brightness would go down with the inverse square of the distance—Newton's famous "inverse square law for light." Thus, if two supernovae had the same brightness at source, and one were twice as far away as the other, then the first would appear one-quarter (= one-half squared) as bright as the nearer one. If Kirshner finds any deviation from the inverse square law, then he will know that space is not flat but curved. The amount of the deviation will give him a measure of the degree of curvature.

For his instruments to stand any chance of detecting a deviation, Kirshner has to examine vast tracts of space, collecting light from one-third of the way across the universe. The light he examines comes from supernovae that had their month of bright, cosmic fame billions of years before the earth was formed.

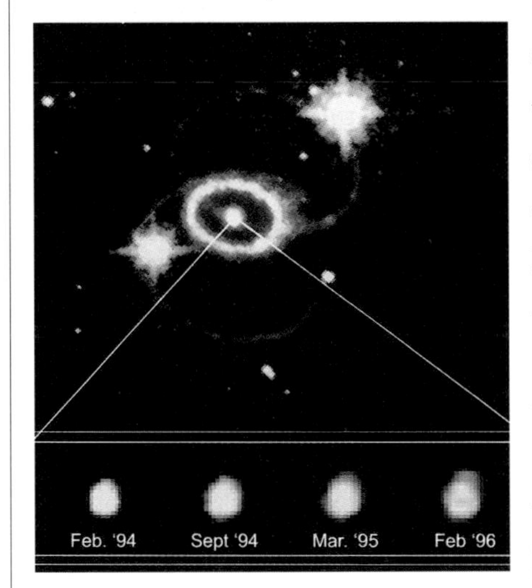

This Hubble space telescope picture shows the debris core of a supernova explosion, seven years after the explosion. The four images at the bottom show the spreading of the interior debris during the course of two years. Calculations show that the debris is expanding at nearly six million miles per hour.

Confident that his approach will work, he predicts a preliminary answer in about two years. He does not see the mathematics as limiting his progress, but the capacity of today's telescopes. "We're at the stage of the Greeks' understanding the shape of the earth," he observes, "because we're working on things that are at the limit of today's technology."

SPACES OF THE MIND

By using mathematics, cartographers can draw maps that enable sailors and airline pilots to navigate the globe, maps that show politicians and relief agencies where the greatest populations live, maps of parts of the sea floor that no human eye may ever see, maps of mountainous terrain accessible only on foot, and maps of the far reaches of the universe, where we see things not as they are today but as they were billions of years before the earth was formed. Computer engineers can take the cartographers' map data and create virtual-reality flight simulators that enable pilots to train without risk of a crash. Satellite data can be used to provide motorists with computerized maps in their cars, pinpointing the vehicle's position at any moment in time.

The key to all these advances is the idea of a map: a representation of the "lay of the land," whatever "land" that might be. And the key to maps is mathematics. All maps are drawn using mathematics.

But mathematics can also be used to draw maps of a different terrain—a terrain you won't find in the physical universe. Mathematics can be used to explore, measure, and map an abstract, mental universe: the universe of mathematics itself.

This picture from the Hubble telescope shows a small portion of a supernova remnant fifteen thousand years after the supernova explosion.

Physicists are not certain whether real space is curved or flat. It has to be one or the other, of course. But a physicist cannot choose to study, say, flat space; the physicist studies real space, the way it is, whether or not it is ultimately observed to be flat or curved.

A mathematician, on the other hand, can study flat space or curved space, according to interest. This is because the mathematical universe is an abstract idealization of the real universe, a what-if idealization where the mathematician can postulate, say, a particular curvature, and then investigate the properties of such a space. Or maybe the mathematician decides to investigate a space of four or more dimensions. Again, such freedom is not available to the physicist, but there is nothing to prevent the mathematician from heading off—mathematically speaking—into four dimensions.

Until recently, there were only two ways to carry out such explorations. Either the mathematician spent a lot of time drawing intricate pictures, as was the case with some talented German geometers in the late nineteenth and early twentieth century, or else the work had to be done entirely in the mind, with the results of the investigations being conveyed to other mathematicians not in pictures but by pages of mathematical formulas. Today, mathematicians have another approach: they can turn to the computer to bring abstract spaces to life in a way that others can share—and indeed experience—including nonmathematicians for whom the formulas have no meaning.

Three possible shapes of the universe as generated by Jeff Weeks on his computer.

Jeff Weeks is one such mathematician. He creates mathematical universes in his computer. Some of his worlds seem quite familiar, but others are quite bizarre. One of his more bizarre worlds is a closed room—though it doesn't appear bizarre until you enter it. At first it looks just like an ordinary room you would find in any house. But in Weeks's room, you can pass through the walls, or the ceiling, or the floor. That's already a bit unusual. But something really strange happens when you try to pass through a wall, ceiling, or floor to leave. If you exit the room by passing through one wall, you find you reenter by the opposite wall. If you leave through the ceiling, you find you are reentering through the floor.

If you leave through the floor, you find you are reentering through the ceiling. Weeks's computer world is a modern-day Dante's hell: though you can pass through the walls, ceiling, and floor, you can never leave the room. Once inside, you must remain there forever—at least until you turn off the computer.

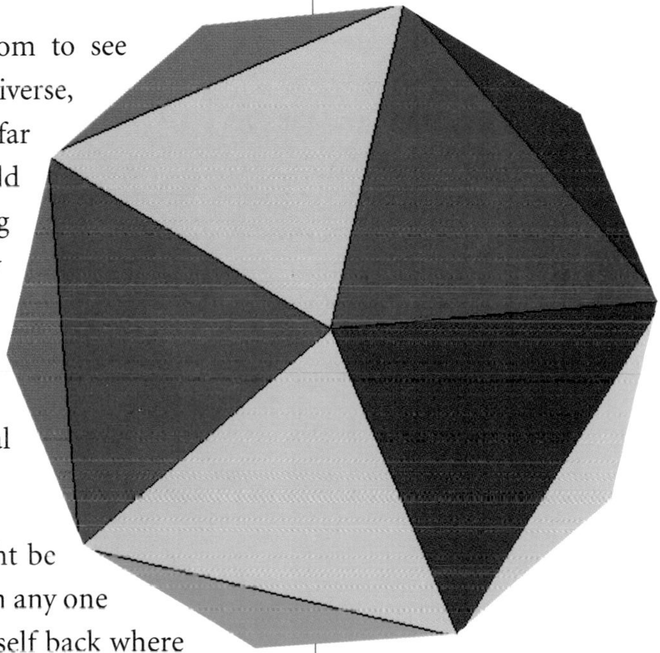

Weeks created his bizarre, no-escape room to see what it would be like to live in a closed universe, a universe where, if you were to travel far enough in any one direction, you would eventually come back to your starting point. In many ways, the result is very much like a computer game, a virtual-reality experience he can share with his young son. Then again, Weeks's computer world might just be a model of the real universe we live in.

Space might very well be "closed." It might be that if you were able to travel far enough in any one direction, you would eventually find yourself back where you started. Of course, because of the huge distances involved, such a journey would be a physical impossibility. But that's not the point. The aim of researchers like Robert Kirshner, who are trying to discover the curvature of space, is not to establish some cosmic travel agency; rather, they want to understand the universe we live in.

It may be that we will never be able to transport our bodies beyond the confines of our own solar system. But with the aid of telescopes, we can embark on a mathematical voyage of discovery. Weeks's computer worlds give us a foretaste of what we might find when we are able to make such a mathematical journey of discovery.

In the closed-room universe—with two star shapes included—the portion of a spaceship that has exited the universe at the right side is already entering the universe again on the left side. There is nowhere else for the spaceship to go.

Like many mathematicians, Weeks is very much a modern-day explorer. But for Weeks, exploration does not require months of planning and preparation. After a day spent in the real world, he can sit down in his room in the evening, turn on his computer, and start to explore a very different world: the mathematical universe. His body remains in his physical room; his mind is transported into the abstract, mathematical world created in his computer.

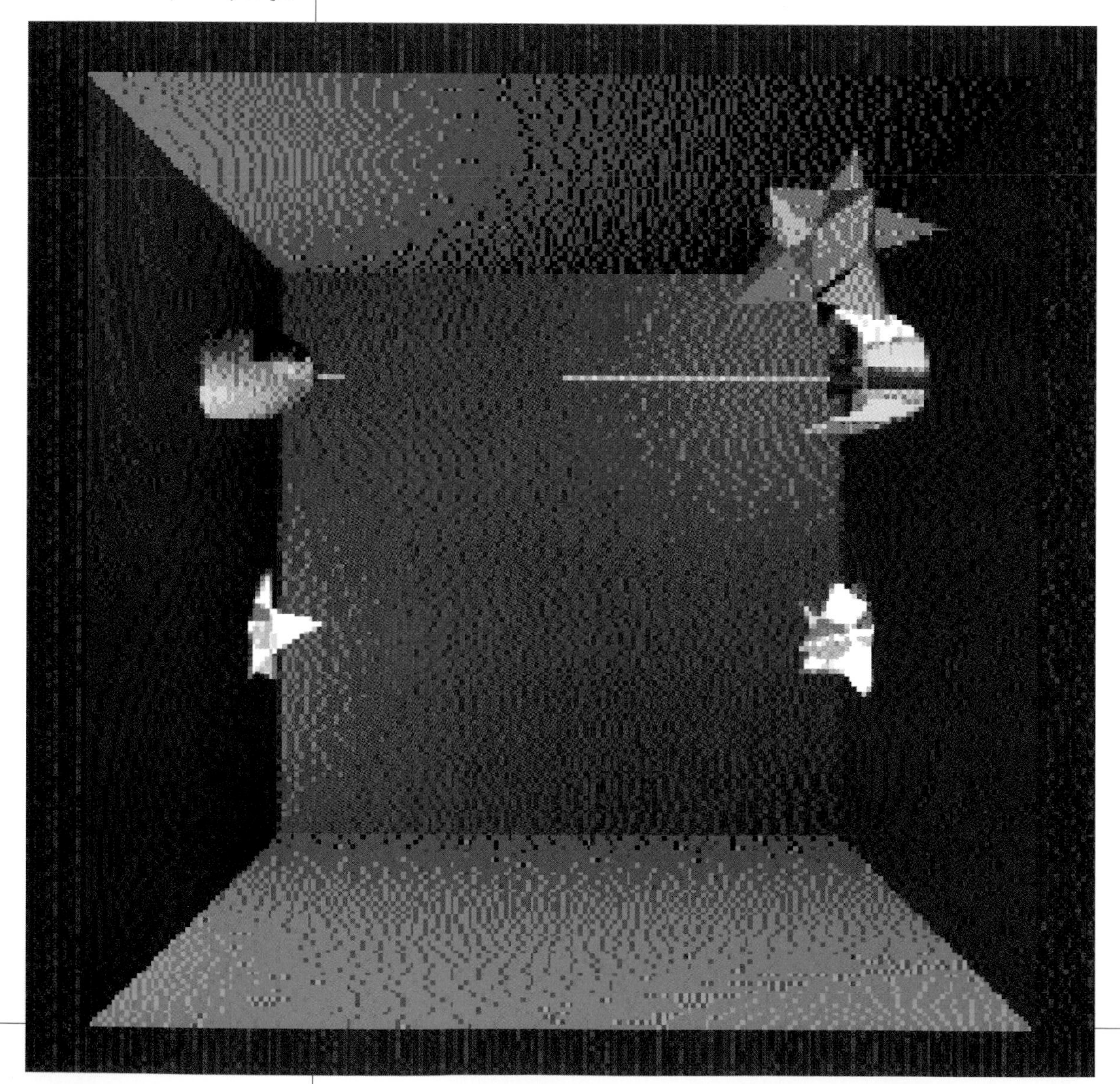

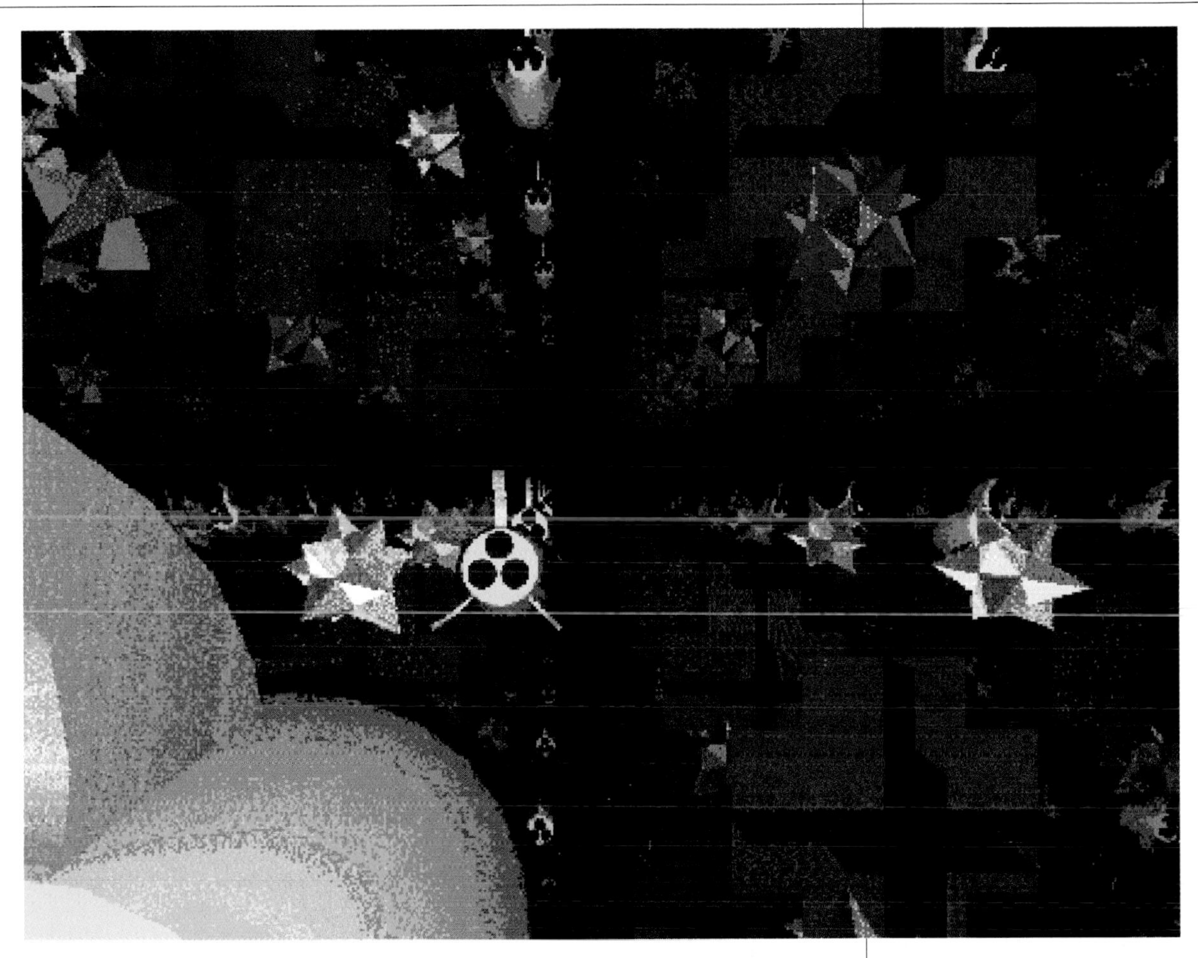

If the walls of the closed room are made transparent, the pilot's line of sight will not stop at the wall in front of him, but will "wrap around" the room so that he will see an image of the back of his own spaceship. But his line of sight will also wrap around the room to his right and left sides, and above and below him, so he will be surrounded by an array of infinitely repeating images, as shown here.

"Real nature is beautiful," Weeks remarks. "You can go out and look at a flower or look off into space. It's a miracle that we have this physical universe to explore. I think it's no less a miracle that we also have this mathematical universe to explore, a universe full of beautiful, intricate patterns lying there, waiting to be discovered."

Chapter 6

CHANCES OF A LIFETIME

Las Vegas—acknowledged capital of the United States gambling industry. Every year thirty million people flock to this small town in the middle of the Nevada desert to try their luck. Some will come to bet thousands of dollars; others will wager no more than fifty. Big spenders or small punters, they are all betting—literally—on Lady Luck being in their favor.

Chance events being what they are, some of them will indeed leave Las Vegas with more money than they had when they arrived. But most won't. They can't. The mathematics of gambling tells us they can't. That mathematics has been around since the seventeenth century, but from the tense activity to be seen every day in casinos not just in Las Vegas but all over the world, it would appear that most people are not aware of it—or else choose to ignore it.

There is one and only one way to win at roulette: own the casino. The mathematics of roulette guarantees that the person who owns the roulette wheel will always come out ahead—not in every single game, but most likely over the course of an entire evening and certainly over a period of a week or more.

Mathematician Ed Packel has made a detailed study of games of chance, including roulette. He has no doubt about your chances: "The odds of winning in a casino are very, very small. In general, with any game, the odds are stacked against you. That does not mean no one will win, but it does mean that any individual who plays for a reasonably long time is going to lose."

The bright allure of the Las Vegas strip at night.

"Each game has its own specific odds," Packel explains. "Probably the most predictable of the games is roulette, where for every bet you make, you can expect, in a mathematical sense, to lose approximately five and a quarter cents on every dollar you bet."

Packel is not talking about a crooked roulette wheel, rigged by the casino to cheat the customers. Even with an "honest" wheel, the odds are still stacked in the casino's favor. The mathematics is surprisingly simple. The roulette wheel is divided into 38 equal compartments. The chances of the ball falling into any one of the 38 compartments are equally likely. So the odds against the ball falling into the compartment you choose are 37 to 1. However, the casino pays odds of 35 to 1. (The odds offered by the casino on the various other kinds of bets that can be made in roulette are likewise figured to give the house a small but reliable edge.) The difference between the odds against your winning and the payout odds offered by the casino is called the "house advantage." It ensures that over the course of many games, the casino will make a steady profit.

And so it is with all the games offered by the casino to try to part you from your money. Packel adds: "Every game is carefully constructed so that if you do the mathematics, it turns out that the casino will have a favorable expectation of winning."

Of course, many people who are aware that the odds are stacked against them choose to gamble anyway. They enjoy the excitement caused by the uncertainty and, as they say, "If you don't play, you don't have any chance of winning." But the people in the casinos who are definitely not gambling are the casino owners.

They don't need to take any chances. They rely on the mathematics, and they know it won't let them down.

You want to see the power of mathematics? In America today, gambling is a $40 billion business that attracts more customers than baseball or the movies. And it is growing faster than any other industry. By using mathematics, casinos arrange the games so that they make an average of three cents on every dollar. As a result, they enjoy annual profits of $16 billion.

State lotteries stack the odds even more in their favor, keeping almost fifty cents of every dollar wagered. The result is $10 billion a year for public funding.

And it all rests upon a series of letters exchanged by two French mathematicians in the middle of the seventeenth century.

FIGURING THE ODDS

Games of chance are as old as society itself. As early as 3500 B.C., bets were placed on the roll of dice. Those early dice consisted of the small, squarish knuckle bone, or astragalus, taken from the ankles of sheep or deer. Paintings of such games have been found on the walls of Egyptian tombs and on Greek vases, and polished astragali have been found in archaeological excavations in many parts of the ancient world.

Chance, it seems, has always held a fascination. Indeed, according to ancient Greek mythology, the modern world began when the three brothers Zeus, Poseidon, and Hades rolled dice for the universe. On that occasion, so the story goes, Zeus won the first prize, the heavens, Poseidon took second prize, the seas, and Hades had to settle for hell.

Oddly, despite the prevalence of gambling since ancient times, it was not until 1654 that anyone worked out the mathematics of games of chance. This might seem particularly surprising in the case of the Greeks, for whom mathematics was the supreme form of knowledge. Plato regarded mathematics as the key to all knowledge, and Aristotle sought to use mathematics to understand the heavens (astronomy), the earth (geometry), language and thought (logic), and music (the theory of scales). Why, then, did the ancient Greeks, who sought perfect order in so many things, not try to find a mathematical order in the roll of a die?

Almost certainly, the answer is that they did not believe there was any order to be found—"chance" was, to them, the complete absence of order. Aristotle wrote: "It is evidently equally foolish to accept probable reasoning from a mathematician and to demand from a rhetorician demonstrative proofs."

This ancient Roman wall drawing depicts women playing a game of dice with dice made of astragali.

In a sense, the Greeks were right. There is no order in a purely chance event. The outcome is entirely unpredictable. The key to finding a hidden order in chance—a mathematical pattern—was to look not at a single chance event but at what happens when that same event is repeated many times. When a chance action is repeated many times, an ordered pattern emerges—an ordered pattern that can be studied using mathematics.

Assuming the dice were not loaded, had Zeus, Poseidon, and Hades agreed to come back and repeat their wager every year, then over the centuries each one of them would have had to endure hell for only one-third of the time, with the other two-thirds split evenly between the heavens and the oceans.

Dice were what interested a sixteenth-century Italian physician called Girolamo Cardano. When he was not tending the sick, Cardano could be found either at the gaming tables or engaged in mathematics. Combining these two passions, he made the first step toward a mathematical theory of chance by showing how to assign numerical values to the possible outcomes of a roll of a die. He wrote up his observations in a book titled *Book on Games of Chance,* first published in 1525, then revised by the author in 1565.

Suppose you roll a die, said Cardano. Assuming the die is "honest," there is an equal chance that it will land face up on any of the numbers 1 to 6. Thus, the chance that each of the numbers 1 to 6 comes face up is 1 in 6, or 1/6.

Today, we use the word *probability.* We say that the probability that the number 5, say, is thrown is 1/6. The probability of throwing either a 1 or a 2, Cardano reasoned, must be 2/6, or 1/3, since the desired outcome is one of two possibilities from a total of six.

This Renaissance ceiling fresco by Luigi Sabatelli depicts the pantheon of the Greek gods, on the mythical Mount Olympus. In Greek mythology, the fate of mere mortals is continually disrupted at the whim of the often ill-tempered, all-too-human gods.

Cardano went further, calculating the probabilities of certain outcomes when the die is thrown repeatedly, or when two dice are thrown at once. For instance, he reasoned that the probability of throwing, say, a 6 twice in two successive rolls of a die is 1/6 times 1/6, that is, 1/36. You multiply the two probabilities since each of the six possible outcomes on the first roll can occur with each of the six possible outcomes on the second roll, that is, thirty-six possible combinations in all. Likewise, the probability of throwing, say, a 1 or a 2 twice in two successive rolls is 1/3 times 1/3, namely, 1/9.

For a pair of dice, what is the probability that the two numbers thrown will add up to, say, 5? Here is how Cardano analyzed that problem. For each die, there are six possible outcomes. So there are thirty-six (6×6) possible outcomes when the two dice are thrown: each of the six possible outcomes for one of the two dice can occur with each of the six possible outcomes for the other. How many of these outcomes sum to 5? List them all: 1 and 4, 2 and 3, 3 and 2, 4 and 1. That's four possibilities altogether. So of the thirty-six possible outcomes, four give a sum of 5. So, the probability of a sum of 5 is 4/36, that is, 1/9.

With Cardano's analysis, a prudent gambler might be able to bet wisely on the throw of the dice—or perhaps be wise enough not to play at all. But Cardano stopped just short of the key step that leads to the modern theory of probability. So too did the great Italian physicist Galileo, who rediscovered much of Cardano's analysis early in the seventeenth century, at the request of his patron, the grand duke of Tuscany, who wanted to improve his performance at the gaming tables.

Girolamo Cardano.

Both Cardano and Galileo concentrated on assigning numerical values to the outcomes of chance events such as a roll of a die. They did not look beyond the gaming table to ask themselves if there was a way to use numbers—probabilities—to assess outcomes in a more general setting. It was when mathematicians addressed that question that humanity took a major leap toward the present era of professional risk management.

In 1654, the French mathematicians Blaise Pascal and

"Probability theory helps you to put some kind of pattern on randomness. You can't tell what's going to happen from step to step, but you can predict what's going to happen over the long haul."

ED PACKEL
mathematician and gambling expert

Pierre de Fermat exchanged a series of letters that most people today agree was the beginning of the modern theory of probability. Though their analysis was phrased in terms of a specific problem about gambling, the two mathematicians looked beyond the problem itself and developed a general theory that could be applied in a wide variety of circumstances—applied to predict the likely outcomes of various courses of events.

The problem that Pascal and Fermat examined in their letters had been around for at least two hundred years: How do two gamblers split the pot if their game is interrupted halfway through? For instance, suppose the two gamblers are playing a best-out-of-five dice game. In the middle of the game, with one player leading two to one, they have to abandon the game. How should they divide the pot?

If the game were tied, there wouldn't be a problem. They could simply split the pot in half. But in the case being examined, the game is not tied. To be fair, they need to divide the pot to reflect the two-to-one advantage that one player has over the other. They somehow have to figure out what would most likely have happened had the game been allowed to continue. In other words, they have to be able to look into the future—or in this case, a hypothetical future that never came to pass.

Blaise Pascal.

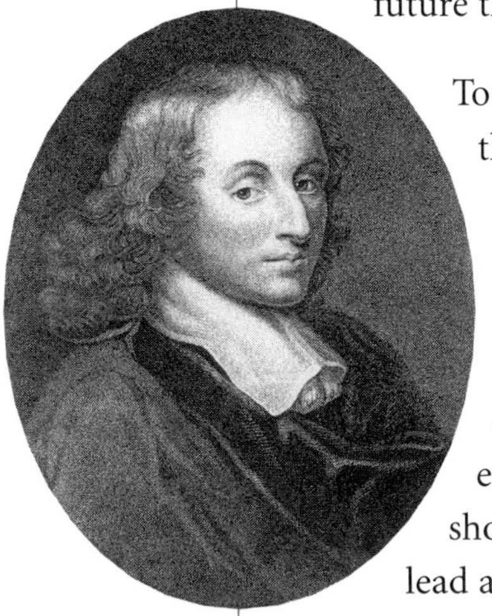

To arrive at an answer, Pascal and Fermat examined all the possible ways the game could have continued, and observed which player won in each case. In the case of the best-of-five dice game that is stopped after the third round with one player in the lead by two to one, there are four possible ways the game can be completed. Of those four, three are won by the player in the lead after the third round. So the two players should split the pot with 3/4 going to the person in the lead and 1/4 going to the other.

Here is the reasoning Pascal and Fermat used. Suppose Pascal and Fermat are themselves the two players, and the first three throws result in wins for Pascal, then Fermat, then Pascal. Thus, Pascal leads two games to one when the game is interrupted. If the final two throws had been made, the possible outcomes for those two throws are wins for Pascal–Pascal, for Pascal–Fermat, for Fermat–Pascal, or for Fermat–Fermat. In three of these four cases, Pascal wins the series; only in the last case does Fermat win the series. Of course, in practice, in either of the first two combinations, the players would probably stop playing after the first additional throw (the fourth in the series), since they would realize that with three wins to his credit altogether, Pascal would have won the series. But in order to obtain an accurate mathematical look into the "what-if future," you have to consider all possible plays of all five games.

Pierre de Fermat.

By showing how to examine all future courses of events in order to assign numerical likelihood values to future possible events, Pascal and Fermat did far more than put to rest a long-standing brainteaser. They opened the door to the present era of risk assessment and risk management.

WHEN THE ODDS REALLY COUNT

Today, we all live in the legacy of Pascal and Fermat. We constantly try to predict the future. The weather forecasters tell us the chances that it will rain tomorrow, and on the basis of that information we decide whether to take an umbrella when we go out. Stock market investments are placed according to the probabilities of future performances of different companies. We take out insurance to protect ourselves from the outcomes of unpleasant events. We never know when we might find ourselves having to figure some really important odds.

Crash-test dummies await their turn to help study the ways seat belts and air bags protect us.

But there is a problem. When we leave the gambling tables and try to figure out odds in everyday life, we find that it is not possible to determine those odds using pure reasoning, as Cardano did for throwing dice and Pascal and Fermat did for the unfinished game. There is no roulette wheel for everyday life. Where the real world is concerned, we have to go out and collect data. We enter the world of statistics.

Want to know the probability that you will die due to a tobacco-related cause? There is no purely mathematical way to work this out. But by collecting data, statisticians have been able to provide an answer. They have discovered that each year four hundred thousand Americans die because of tobacco—this is equivalent to three fairly full jumbo jets crashing every day and killing all on board.

Want to know how your chances of surviving an automobile crash are affected by wearing a seat belt? Again, you can't figure this out using mathematics alone, but by collecting data, statisticians have again come up with an answer. Wearing a seat belt cuts your chances of dying in a car crash by 50 percent. Air bags add another 11 percent of protection.

These days, statistics is big business. In many ways, statistics rules our lives. What we can buy in the store, what automobiles are available to buy, what we can earn, what movies we can choose to see, what advertisements we see on television—these and many other things in our lives are heavily influenced by statistics. Statisticians collect numerical data and then they use those numbers to understand how we live, what our needs and wants are, even how we are likely to die. They look at the data and try to draw conclusions. Statistician George Cobb describes statistics as looking for meaning in numbers the way we search for meaning in a play or a novel. "Statistics involves finding an interplay between pattern and numbers and their meaning," he says.

Mathematician Hal Stern teaches probability and statistics at Iowa State University. A lifelong baseball fan, he uses his mathematical knowledge to chart the progress of his local team, the Iowa Cubs. "I always liked numbers," he declares. "I was the guy on the Little League team who computed people's batting averages and the like."

"Statistics is numbers that are part of a story."
GEORGE COBB
statistician

Baseball managers use probability theory all the time to decide the optimum strategy, Stern explains, even if they don't think of it that way. "Deciding whether to bunt or who to play on a particular day depends on the manager's ability to assess probability. The decision making is hard, because you don't know what's going to happen. You make decisions based on what is going to give you the best chance of success."

Sometimes, trying to assess what will happen in the future takes on far more significance. For example, when faced with a patient having a potentially fatal illness, for which the only known treatment itself carries a risk of death, a doctor has to compare the probability of success of the treatment with that of the patient surviving the illness unaided. In such cases, probability theory can make the difference between life and death.

Baseball is a game of split-second decisions. Managers keep elaborate statistics to help make decisions, such as what the chances are that a runner will reach home plate before a catcher can tag him out.

TO IMMUNIZE OR NOT TO IMMUNIZE

The example just given, of a doctor having to decide whether or not to treat a patient, is just one instance of how modern medicine relies on the mathematics of probability—on figuring the odds. Every year, the United States Food and Drug Administration, the FDA, relies upon probability theory to evaluate new treatments for disease.

A new drug is found to work in the laboratory. The FDA must determine whether—and when—to release it for public use. It is not an easy decision, and it cannot be taken lightly. Before the FDA will authorize general use, a properly conducted study has to be made. That's where the mathematics comes in.

Susan Ellenberg is a division director at the FDA. Her department evaluates the effectiveness of new vaccines. She uses the example of the polio vaccine to illustrate the evaluation process.

Between 1916 and the 1950s, polio, a crippling and often fatal disease, claimed hundreds of thousands of victims in the United States and around the world. Many of the victims were children. No one knew how the disease was spread. People were frightened. "Everyone was fearful about polio," Ellenberg recalls. "People wouldn't let their kids go swimming. It was one of the most fearful diseases."

Researchers all over the country set to work trying to find a vaccine that would protect against the disease. Finally, Dr. Jonas Salk developed a promising vaccine that seemed safe and effective in laboratory trials. The question then was, Would the vaccine be equally safe and effective outside the laboratory?

If the vaccine were simply released for public use and the incidence of polio declined, how would we know whether it was the vaccine that had caused the decline? It might have been pure coincidence. After all, like most infectious diseases, polio appeared in cycles—some years there were many cases, other years just a few. The problem then was to eliminate the possibility of coincidence and chance.

"People generally think of the numbers and less of what is to me the interesting part: using the numbers to make decisions about everything under the sun."
HAL STERN
mathematician

In such circumstances, the idea is to use what we know about probability in order to eliminate any effects of chance. To test the new polio vaccine, a controlled trial was set up, involving two groups of children. One group would receive the new vaccine; the other group would be given a placebo, a completely neutral substance containing no medication, such as a saline solution. "You needed to be able to say that the group that got the vaccine had fewer cases of polio than the group that didn't get the vaccine," Ellenberg explains.

The trial was conducted in 1954 and involved four hundred thousand schoolchildren chosen randomly from the population. The children were split up into two equally sized groups. Care was taken to ensure that the two groups were very similar. For instance, it would be no good if the children in one group came from wealthier homes than those in the other, or if the two groups were from different parts of the country. In the former case, standard of living might make a difference; in the latter case, maybe location was a factor.

Because the testers were confident that the two test groups were alike in all respects, when the results came in and showed a much lower incidence of polio among the group that had received the vaccine, the conclusion was clear: the Salk vaccine worked. General use of the vaccine was authorized.

Sometimes the use of clinical trials is criticized. After all, in the case of the trials for the Salk vaccine, some of the children who were put into the placebo group contracted polio. Had they been "lucky" enough to be assigned to the other group, they would have received the vaccine and might well have resisted the disease. But when there are so many factors that might play a role, there is no other way. As Ellenberg says, "Clinical trials are necessary because that's the only way we can determine whether a new treatment is safe and effective for its intended use."

To emphasize her point, Ellenberg cites another trial carried out a few years ago. "Several years ago, a new treatment was being developed for what is called Lou Gehrig's disease, amyotrophic lateral sclerosis. This is a disease for which there hadn't been a standard treatment. People lose their muscle function and eventually die."

The new treatment proved effective in the laboratory, and many called for its immediate release. The FDA insisted on a clinical trial. Ellenberg continues her account: "This trial did go forward as a placebo-controlled trial, and it turned out that the people who got

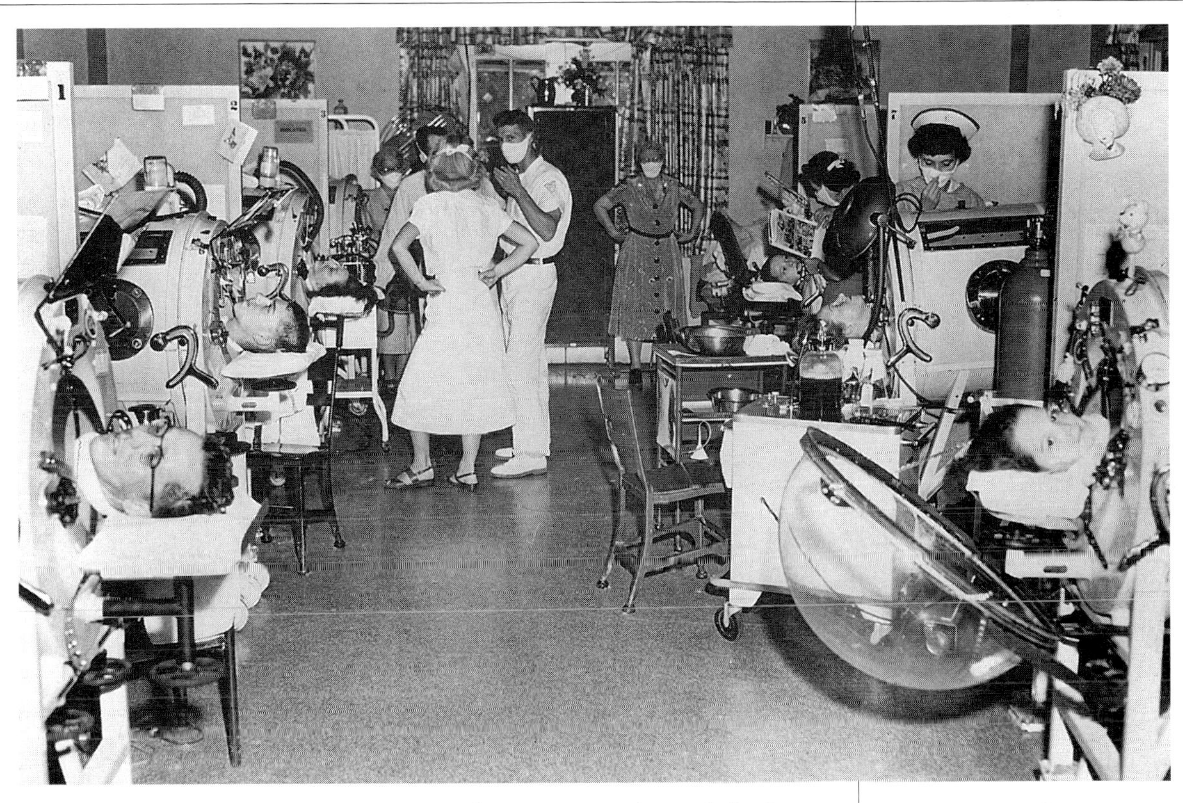

In 1953, polio patients stricken with the disease by the bad luck of chance get treatment with iron lungs.

the new product did worse. The trial was stopped. And that's not going to be a treatment that's going to be widely available. You can't predict in advance."

ORDER IN THE CHAOS

In a clinical trial, with the possibility of chance events playing their hand, the only way to proceed is to use our understanding of probability to fight the effects of probability. For instance, it is crucially important that the trial group is selected randomly, so that anyone in the target group (children, women, whoever) is equally likely to be chosen.

Likewise, the division of the trial group into two—the test group who will receive the medication and the control group who will be given a placebo—must be made randomly. The testers must be confident that the only thing that distinguishes the two groups is that one receives the medication, the other does not. Not even the doctors who administer the samples are allowed to know whether they are giving the patient a drug or a completely neutral placebo.

The use of random assignment is so effective because, in the long run, randomness is extremely reliable. Surprising though it may be, the complete disorder and impossibility of prediction for a single random event gives rise to a highly regular and predictable pattern when a large number of similar events occur.

One of the first people to study this phenomenon was John Galton, an amateur mathematician who lived in the nineteenth century. A device called a Galton board, named after him, can be used to illustrate the order that emerges from a large number of random events.

A Galton board consists of a thin sandwich of wood and glass into which small balls (or coins) can be dropped. The path of the balls is impeded by a regular array of pins (or pegs). When a ball hits a pin, it bounces to the left or the right with equal probability. Then it hits another pin, and again bounces left or right. And so on, until the ball reaches the bottom. So, you have a series of perfectly random events, where the ball bounces left or right with equal probability.

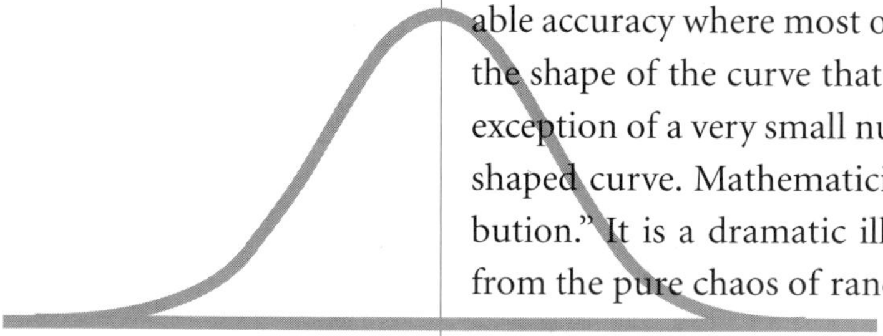

A standard distribution bell curve.

Drop in one ball, and you have no idea where it will end up. But drop in a large number of balls, and you can predict with remarkable accuracy where most of them will land. In fact, you can predict the shape of the curve that the pile of balls will trace out. With the exception of a very small number of balls, you will always get a bell-shaped curve. Mathematicians call the shape the "binomial distribution." It is a dramatic illustration of the order that can emerge from the pure chaos of random events.

In today's information-rich and prediction-loving society, random selection has become so important that it is now done routinely by computers. These days, it is rare for anyone in the business of evaluating drugs or testing public opinion not to be aware of the importance of using a genuinely random sample.

It was not always so. In the U.S. presidential election of 1936, Republican Alf Landon was running against the incumbent president, Franklin Delano Roosevelt, a Democrat. In order to try to predict the outcome, the *Literary Digest* magazine conducted the biggest election poll ever, sending out over ten million ballots. When the results came in, they predicted a landslide victory for Landon.

During the same election, a pollster named George Gallup also conducted a poll. His survey was much smaller than that run by the *Literary Digest.* He questioned a mere fifty thousand voters, on the basis of which he predicted a victory for Roosevelt.

Roosevelt won. Gallup, with his poll of fifty thousand,

	Black	White
First Sample	49	51
Second "	44	56
Third "	50	50
Fourth "	52	48
Fifth "	47	53

Famous polling expert George Gallup demonstrates the principle of sampling. The bowl at left contains 500 white and 500 black beans, thoroughly mixed. With eyes closed, Gallup proceeds to pick a total of 100 beans and will find, when he is done, that there will be approximately 50 white and 50 black beans in the sample he has selected. The chart on the desk shows the exact tallies of white and black beans from five successive runs of this experiment.

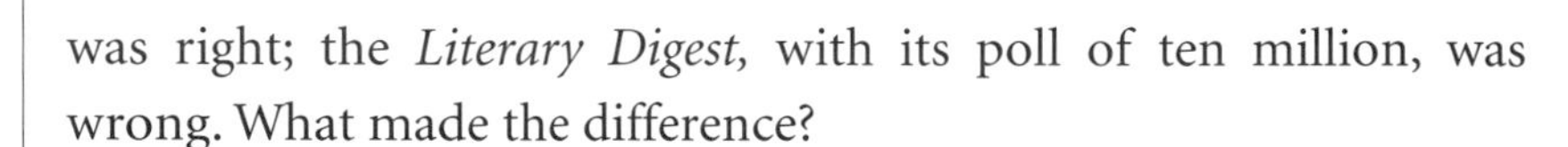

was right; the *Literary Digest,* with its poll of ten million, was wrong. What made the difference?

Randomness. Gallup selected his fifty thousand voters at random from the entire population and went out and asked them all how they would vote. The *Literary Digest* relied on its own subscriber list, club membership lists, and telephone books. As a result, its sample was anything but random. In particular, it excluded most of the poor, who were unlikely to subscribe to *Literary Digest,* belong to clubs, or, in those days, have a telephone.

The cover of the issue of *Literary Digest* that predicted President Franklin Roosevelt would be defeated in a landslide by Republican candidate Alf Landon.

Another problem for the *Literary Digest* poll was the response rate. It was a mail-in poll, and only 20 percent responded. For the most part, they were the people who were angry with Roosevelt and who wanted to see him replaced by Landon. By and large, the people who were content with Roosevelt did not bother to reply to the poll request.

What does it take to have a reliable poll? Bill Kaigh can provide the answer. Kaigh is a polling expert and a professor of mathematics at the University of Texas in El Paso. He and his wife, Geri, constitute Kaigh Associates, a public-opinion polling business that has operated in the El Paso region for the past ten years.

"What are the ingredients of a good poll?" Kaigh echoes. "The very first question you need to ask is, Is the sample scientifically selected? Is there randomization involved? And this gets down to the question, Did the survey select the sample, or did the sample select the survey?"

In the *Literary Digest* poll, the sample selected the survey: only the people with strong anti-Roosevelt feelings responded. For the same reason, the call-in polls you often see on television (often using 900 numbers) are not reliable. For a poll to be accurate, the sample must be chosen randomly from the larger population.

Kaigh was recently asked to organize a poll by the El Paso *Times* and the local Channel 7 television station, to predict the likely outcome of an upcoming mayoral election. He began with the list of registered El Paso voters. Then he used a computer to randomly select a group of three hundred. Because his selection was made at random, Kaigh was confident that a sample of just three hundred was sufficient. "A lot of people have a very difficult time conceiving that by taking a relatively small sample, a few hundred, you can actually make generalizations about a population of perhaps even millions," Kaigh observes. Randomness is the key.

"Being able to handle uncertainty—that's basically what probability is, what statistics is about. That's always been very fascinating to me: being able to foretell the future."
BILL KAIGH
mathematician and public-opinion pollster

Having selected the polling sample, Kaigh's next step was to develop the questionnaire. This too has to be done scientifically. "We'll go through a series of questionnaire revisions," Kaigh explains. "We don't want to use emotional terms that might lead the respondent or bias a question." For example, Kaigh avoids using a title such as "current mayor," since that might lend authority to the incumbent over the opposing candidates.

Finally, after some pilot testing, the poll was ready to use. Working from her sample list, Geri Kaigh started to make her calls.

"Hello, I'm calling from Kaigh Associates," she began. "We're conducting a scientific poll." Moving steadily through her list of carefully prepared questions, she was careful to give a neutral inflection so as not to lead the respondent. It was a slow process, but steadily, the data were accumulated.

A week before the election, Bill Kaigh sat down to analyze the results. On that occasion, the outcome was clear: 71 percent of the sample declared their intention to vote for the incumbent. Because the sample had been chosen randomly from the entire voting population, Kaigh was very confident in the prediction. When election day came, his prediction turned out to be correct.

In a democracy and a free-market economy like the United States, public-opinion polling has become increasingly significant in all our lives, Kaigh observes. "We're interested in knowing what the majority of people feel. That's behind our basic political structure. New products are typically tested using surveying techniques. Public-opinion polling determines a lot of what is seen on TV. Virtually every facet of our everyday lives is, to some extent, influenced through public opinion."

This example of a computer graphic presents the result of a public-opinion poll measuring the level of personal satisfaction all around the world.

THROUGH MATHEMATICS, PEACE OF MIND

Many people who have never set foot in a casino, and may even declare themselves against gambling, will nevertheless regularly place bets based upon the value they put on their lives, their house, their car, and other possessions. For that is exactly what you are doing when you take out insurance. The insurance company estimates the chances that, say, your car will sustain serious damage in an automobile accident, and then offers you odds against that happening. If there is no accident, the insurer keeps the relatively small premium you have paid. If there is an accident, and the car is a write-off, the insurer pays out the cost of a new car.

To offset the huge difference between the possible payout and the premium, the insurer uses mathematics. Based on the measured (or estimated) frequency of accidents, the insurer sets the premiums for the policies sold so that the total amount received in premiums exceeds the total amount likely to be paid out. Once the premiums have been set and the policies sold, the insurer is in the hands of Lady Luck. In a "bad" year (for both the insurer and the insured), there will be more claims than anticipated, and the insurer will have to pay out more than usual. Profits will fall—the company might even take a loss. In a "good" year, there will be fewer claims than expected, and the company's profits will be high.

Life insurance policies, for example, are based on life expectancy tables, which list the number of years a person is likely to live, depending on his or her current age, place of residence, occupation, lifestyle, and so on.

Life expectancy tables are drawn up by making a statistical survey of the population. The first such survey was carried out in London in 1662, by a merchant called John Graunt, who made a detailed analysis of the births and deaths in London between 1604 and 1661. His main source of data was the "Bills of Mortality" that the city of London had started to collect in 1603.

It is not exactly clear what inspired Graunt to carry out his study. It might have been pure intellectual curiosity. He wrote that he found "much pleasure in deducing so many abstruse, and unexpected inferences out of these poor despised Bills of Mortality." On the other hand, he also seems to have had a business objective. He wrote that his research enabled him "to know how many people there be of each Sex, State, Age, Religion, Trade, Rank, or Degree, &c. by the knowing whereof Trade and Government may be made more certain, and Regular; for, if men know the People as aforesaid, they might know the consumption they would make, so as Trade might not be hoped for where it is impossible." Whatever his motivation, Graunt's work marked one of the very first examples of modern statistical sampling and market research.

Thirty years after Graunt published his finding, the famous English astronomer Edmund Halley—he of Halley's comet—carried out a similar, though much more intensive, analysis of mortality figures. Halley's data came from the town of Breslau, in Germany, the present-day Wroclaw in Poland. Halley's interest was purely scientific. The Breslau data that was made available to him was very detailed and accurate, collected on a monthly basis between 1687 and 1691, and Halley wondered what he could make of it.

The answer is, he made a lot of it. Halley's mathematical analysis was so comprehensive that a wise insurer could have used his methods as the basis for a profitable life insurance business. Insurers at the time were, however, not sufficiently wise, and indeed it was to be over a hundred years before they would start to base their policies on reliable data and analysis.

The modern insurance business started to develop toward the end of the eighteenth century. One of the first of today's international insurance companies, the famous Lloyd's of London, started in 1771, when seventy-nine individual insurers entered into a collaborative agreement. The name they chose for their new company was that of the establishment where they had hitherto carried out their business—mostly shipping insurance: Edward Lloyd's coffeehouse on London's Lombard Street. Lloyd himself had played no small part in the affair: in 1696, five years after he had opened the coffeehouse—and doubtless having observed who his early customers were and wanting to keep them—he started "Lloyd's List," a compilation of up-to-date information on the arrival and departure of ships and the conditions at sea and abroad.

Ironically, the famous insurer Lloyd's of London was itself once the victim of a terrible accident. This print depicts the burning of the Royal Exchange, the building that housed the original Lloyd's offices and that burned to the ground in a raging fire on January 10, 1838.

The first American insurance company—called, appropriately, First American—was a fire insurance company established by Benjamin Franklin in 1752. The first life insurance policies in America were issued by the Presbyterian Ministers' Fund in 1759. The word *policy,* incidentally, comes from the Italian word *polizza,* which means a promise.

The management of risk has become one of the important businesses of the twentieth century. As investment consultant Peter Bernstein writes in his book *Against the Gods,* "The revolutionary idea that defines the boundary between modern times and the past

In the American Midwest, many homeowners take out tornado insurance because of the frequency of tornadoes in this part of the world. In this photograph of a tornado spiral, a house has been sucked up into the center of the funnel.

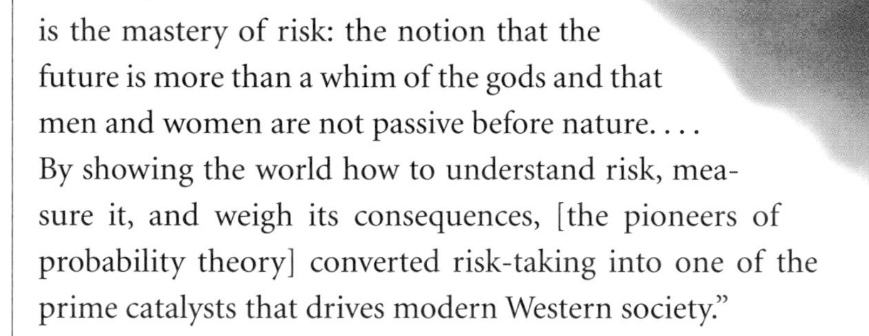

is the mastery of risk: the notion that the future is more than a whim of the gods and that men and women are not passive before nature. . . . By showing the world how to understand risk, measure it, and weigh its consequences, [the pioneers of probability theory] converted risk-taking into one of the prime catalysts that drives modern Western society."

Today, insurers offer policies to cover all kinds of eventualities: death, injury, auto accident, theft, fire, flood, earthquake, tornado, hurricane, accidental damage to household goods, loss of luggage on an airplane, et cetera. Movie actors insure their looks, dancers insure their legs, singers insure their voices. You can even buy insurance to cover something going wrong at your wedding.

Each year, in the United States, there are over two million weddings. According to *Bride* magazine, the average cost is between fifteen and twenty thousand dollars.

When Chad and Adele Smith planned the wedding of their daughter, Rachelle, to her fiancé, Tim, they included insurance for the event, to cover bad weather, illness, photographs not coming out, or anything else that might spoil the day or even prevent the wedding from taking place at all. "Your big investments in life are your home, your cars, and a wedding," Adele explained. "I would think that a wedding ranks right up among those other two things. What else do you do that costs so many thousands of dollars? And to not insure it? You wouldn't think of not insuring your car or your home, so why not the wedding?"

The Smiths approached Fireman's Fund, an insurance company in Marin County, California, that specializes in business and property insurance, but that is also known for insuring unusual things such as sporting events and motion picture productions.

As shown in the photo above of the aftermath of a tornado in a suburban neighborhood, the capricious path of the wildly spinning funnel clouds will often destroy two or three homes and then skip over several others.

"We're trying to make predictions of the future."
GRANT STEER
actuary

Grant Steer has been an actuary at Fireman's Fund for twenty years. "Actuaries are involved in setting rates, trying to decide what the price ought to be for the policies and products we sell," he explains. Steer met with his colleagues to decide what kind of policy to issue to cover weddings such as the Smiths'—something they had not covered before. "Insurance usually has a history of the data from the types of losses that the particular policy they've sold has covered," Steer explains. "So they have some actual statistics on the types of losses they've incurred. Wedding insurance was a new coverage."

So the company had to do its own research, to decide what elements should be insured and what losses to expect. Marketing analyst Sue Lim researched the cost of various elements of a wedding. Should they insure the wedding attire? The wedding photographs? The wedding gifts? She called shops that specialize in wedding clothing to find the typical costs involved. She phoned wedding photographers to find the range of prices they charge. "How often do things go wrong?" she wanted to know.

Steer consulted with his fellow actuaries to try to predict the frequency of wedding cancellation. They observed that the longer an engagement lasts, or the longer a policy is in place before a wedding, the more opportunities there are for things to go wrong. Thus, couples with long engagements represent a greater risk for the insurance company. On the other hand, they didn't think they could charge a lower premium for a short engagement than for a long engagement. So to simplify the rating, they decided to average over the different frequencies or the number of claims per policy that might occur.

"We try to set up a statistical model or a theory as to what the events are going to look like," says Steer. "Our assumptions are that we'll have in general about three claims per hundred. So about three percent of the time we're going to have a claim."

When the data were in, Steer and his colleagues drafted the policy they felt comfortable offering. Because of the low risk, they were able to provide several thousand dollars' worth of insurance for a policy premium of little more than a hundred dollars. It covered wedding cancellation, photographs, wedding attire, wedding gifts, and personal liability.

Steer summarizes the entire process: "We added up as best we could the costs of the various insurable items. The other thing we needed to know was, how often do those losses occur?"

For the company, the mathematics provided the confidence they needed to offer the wedding policy. Provided they had to make a

Hidden behind the veil is the peace of mind provided by the mathematics of insurance.

The photograph of Hurricane Bonnie, opposite, was taken by the Space Shuttle *Endeavour*. The eye of the hurricane is clearly visible at the center of the spiral.

payment on only 3 percent of the policies they issue, they will make a reasonable profit.

Of course, some events are more difficult to predict than others, and among the most notoriously unpredictable are major weather catastrophes and natural disasters, such as hurricanes and earthquakes. To protect themselves against the vagaries of such major disasters, some insurers offer a high-risk form of security called catastrophe bonds. These bonds are sold to high-stakes investors who are betting against a given amount of damage being done by a storm or other disaster over a specified area of population during a given period of time. These investors are, in effect, betting against the weather.

The way that the insurers and investors can estimate the risk involved with catastrophe bonds is by working out the mathematical probability of the kinds of disasters they are hedging against. Analysts compile decades' worth of data about all sorts of natural disasters and calculate the probabilities that given kinds of storms will happen. For example, analysts have established that a storm large enough to cause $1 billion in insurance claims will strike the East Coast of the United States only once in every 100 years, so the probability that the storm would happen during any single year is one in one hundred.

As Grant Steer, the Fireman's Fund actuary, reflects about the business of insuring against risk: "There is no absolute truth. There is no magic roulette wheel to tell us how often a one or a two or a three is going to come up. So we are always having to make estimates from the information we have, from the data we've been able to gather from our commonsense view of the world."

Predicting the future has been a dream of mankind throughout history. Ancient mythology is full of stories of oracles, sages, and

soothsayers having the power to see into the future. Even today, palm reading, crystal gazing, and tea-leaf studying all have their followers. The plain fact is, we do not know how to predict the future. It is probably not predictable in any exact sense. But by using mathematics—the mathematics of probability theory—we can often predict what is likely to happen. We can even assign a numerical value to that likelihood.

Perhaps the most cataclysmic accident we would want to calculate the risks for is a massive meteor impact. The Berringer Meteorite Crater, in northeastern Arizona, which is one-half mile wide, was created fifty thousand years ago when a giant iron meteor smashed into Earth. Scientists have calculated that a meteor of that massive size will hit a land area on Earth once every fifty thousand years, and some experts warn that we may therefore be due for a second catastrophic impact event of this kind.

In that sense, mathematics provides us with eyes to see into the future: one further instance of how the invisible universe we call mathematics allows us to see what would otherwise be invisible.

Chapter 7

A NEW AGE

Today we live in a world that has been created in large part using mathematics—a world of computers, of information technology, and of rapid communications around the world. The creation of that world has presented mathematicians with a new challenge: understand that new world—find *its* hidden patterns—and help

"I think science is really fun. It's like playing in the sandbox. That's what we are doing here at the Media Lab. It's one big sandbox with lots of toys, and we can do anything with it."
PATTIE MAES
computer scientist

Pattie Maes.

One of the places where this kind of investigation is being done is MIT's Media Lab. "Inventing the future" is the phrase that Media Lab scientists often use to describe their research. The vision that lies behind the slogan is of a future in which computation and digital communication play an increasingly significant role. Doing things with computers is what the lab is really about. Media Lab computer scientist Pattie Maes describes it this way: "The main thing we're trying to do at the MIT Media Lab is invent the future and find out how people are going to work and play and learn in the digital future."

For some years now, Maes has been developing a special kind of computer program called a software agent, or "softbot" (for "software robot"). Maes describes a softbot as a "digital butler" that can do your bidding, fetching not food or clothing but information. Software agents are designed to roam the World Wide Web and find information, bringing it back to the user.

With the growth of worldwide computer networks in the late 1980s and early 1990s, culminating in the establishment of the World Wide Web in the mid-1990s, the amount of information that anyone could access from a simple home computer became effectively infinite. There is so much information on today's World Wide Web, available at the cost of a few keystrokes, that the problem is how to find the information you really want among the great mass of information that is available. The Web brought with it both the advantage of an unlimited source of information and the crippling disadvantage of information overload.

The only solution to this digital-age problem of information overload is . . . you guessed it: the very computer technology that caused the problem in the first place. Just as computers can rapidly generate and process vast amounts of information, so too computers can quickly find their way through that information. At least, that was the theory. Maes and others set about putting that theory into practice, developing computer programs that can perform that search-and-sort task.

Tell one of Maes's softbots that you are interested in music of a certain kind and it will produce a list of suggestions of new releases that you might like. Looking for a new dinner recipe to impress that special someone? Send your software agent out onto the World Wide Web to find it for you. Like the perfect butler in a Victorian novel, Maes's softbots are always willing—and able—to do your bidding.

A computer animation of the inside of a computer chip.

Researchers at the MIT Media Lab display their wearable computers.

Maes's research helps to bring the world of information into your computer. Her Media Lab colleague Bradley Rhodes is trying to take your computer out into the world—the physical world of everyday activity. Wearable computers are his interest. With the body of the computer in a small satchel, a specially designed keyboard in one hand, and a small monitor mounted on a hat, wearable computers will probably find their first main use in factories or on building sites, where the user needs to have freedom to move around but also requires access to computer data. Beyond that, no one is really sure what their use will be.

To develop the technology, Rhodes set up his wearable computer to provide him with everyday information at a glance. As he walks around, goes shopping, and so forth, a permanently visible reminder pad that is constantly updated tells him about his next appointment, reminds him to buy milk on the way home, et cetera. Whether people will ever use wearable computers in that way is unclear. A more likely use is by police officers on patrol. As Rhodes observes, "When you take the computer with you, that's when the power really comes into play."

Whether on the desktop or on a wearable computer, at present software agents might act like "butlers," but they certainly don't look like butlers. In fact, they don't look like anything except a computer display. Maes and her colleagues think that people will find it easier to deal with softbots if they can be given an animate personality that you can relate to.

"Once I realized that math was just a tool, it suddenly opened a whole world for me. It allows me to be creative and expressive in whatever domain I want."

PATTIE MAES
computer scientist

"Computers aren't very friendly yet; they are not very personal," Maes admits. "My computer, even though I've been using it for years, still doesn't know me; it doesn't react to me in a different way. So one of the things we're trying to do is make the computer into something more personal, a system that also takes a more active role in trying to help you do whatever you want."

In particular, Maes continues, "Most of the agents that we have don't really have any kind of physical representation on the screen." This is where the Media Lab's project, called ALIVE, comes in. ALIVE stands for Artificial Life Interactive Video Environment. Maes explains: "The ALIVE system is a test bed for trying to build 3-D animated characters that a user can interact with in real time. We have been starting out by modeling simple characters, mainly animals."

An image of Bradley Rhodes as seen through the Photobook Face Database System on a wearable computer screen. This system has the ability to scan the face of a person walking toward you and search a database of eight thousand faces in approximately one second. The system then provides the names and other personal data for the forty closest matches. Possible users include police officers, reporters, and the visually disabled.

Bruce Blumberg is one of the researchers working on ALIVE. He has created

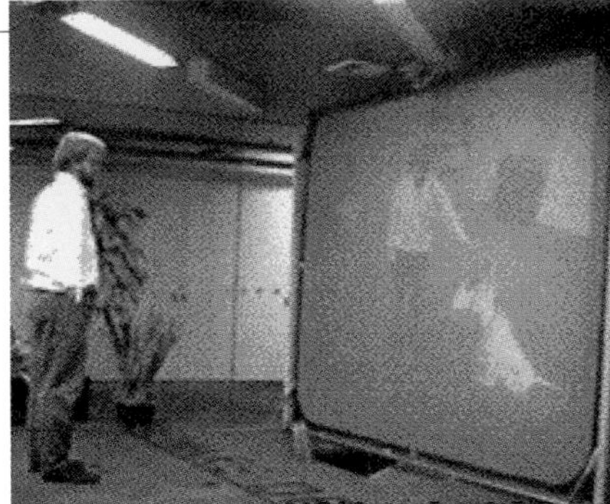

Bruce Blumberg stands in front of the ALIVE screen, above, and interacts with the virtual dog named Silas. The system is based on a magic-mirror metaphor: a user of the ALIVE space sees his or her own image on a large-screen TV as if in a mirror. Computer-programmed animated characters then join the user's image in the reflected world, as seen in the image on the right.

Silas, a software agent designed to appear—and act—like a friendly dog. Inside the research lab, the computer's cameras focus on Blumberg, noting the approximate positions of his hands and feet. On the screen, Silas is programmed to react to Blumberg's motions the same way that a dog would react to its master, responding to commands and moving around in response to Blumberg's movements.

As with Rhodes's experiments with wearable computers, it may be that none of us will ever have software agents that look and act like dogs. The point of the research is to study how people interact with computers and to find out how to make computers easier to use. That's what inventing the future is like. As Maes says, being at the Media Lab is like being in the sandbox.

There is just one admission ticket to the sandbox: mathematics.

Maes adds: "One of the things that makes all of this stuff possible is that pretty much everybody here has some background in mathematics. It doesn't need to be complicated math. We're using mathematics to do something that is really fun."

Blumberg echoes Maes's comment. "Everything you see at the lab is based on computers, so ultimately it relies on mathematics. Because without mathematics, you wouldn't have computers."

THE SOUL OF THE MACHINE

The mathematics that lies behind the computer was developed in the middle of the nineteenth century by an English mathematician called George Boole. In 1854, Boole published his new mathematics in a book titled *The Laws of Thought,* a book that has remained in print ever since.

In his book, Boole showed how to apply algebra to the human mind. Before Boole, when mathematicians did algebra, the letters (or "unknowns") *x, y, z,* and so on generally denoted numbers. In Boole's new algebra, the letters denoted propositions—human thoughts. When you solve an equation in ordinary algebra, the answer is a number. When you solve an equation in Boole's algebra, the answer is another proposition, the logical conclusion to a process of human reasoning. Today, we call the mathematics Boole worked out "Boolean algebra."

George Boole.

The Jevons logic machine has a small keyboard and an arrangement of pulleys and levers inside the cabinet. Each key has a letter or other symbol printed on it and each represents a certain type of logical statement. Propositions or "questions" can be offered to the machine by pressing the right keys; in response, another set of letters and symbols flicks up in the face of the cabinet to present the corresponding pattern of thought.

Startling though it appears at first to think of applying algebra to human thoughts instead of numbers, mathematically the idea is fairly straightforward.

The simplest way to think of algebra is as the science of putting things together. It's a little bit like playing with Lego blocks. When you take a pile of Lego blocks, there are certain rules you have to obey in order to put the blocks together. We can call those rules the algebra of Lego.

Ordinary (numerical) algebra looks at the ways you can put numbers together to form new numbers. You can add them, you can multiply them, you can subtract one from another. Algebra studies the rules for performing those operations on numbers. (For example, one rule is that when you add two numbers, it doesn't matter which number comes first; the answer is the same in both cases.) To describe those rules precisely, you have to use letters x, y, z, and so on to represent arbitrary numbers (or arbitrary Lego blocks!), and that's why when most people think of algebra they think of rules for manipulating symbols.

In the nineteenth century, mathematicians started to realize that they could develop algebra to describe the rules for putting other kinds of things together besides numbers. In particular, Boole developed an algebra for putting together human thoughts to give new thoughts. Taking arithmetic as his model, Boole focused his attention on just three procedures for combining thoughts, corresponding to the arithmetic operations of adding, multiplying, and forming the negative.

First, you can join two thoughts together with the word *or* to form a single new thought. For example, joining the thought "It will rain today" with the thought "It will snow today" gives the new thought "It will rain today or it will snow today." You can also join two thoughts together with the word *and*. For example, joining the thought "Alice will have pizza" with the thought "Bill will have a burger" gives the new thought "Alice will have pizza and Bill will have a burger." Finally, you can take any thought and negate it. For example, negating the thought "I will go by plane" gives the thought "I will not go by plane." Just as we can write down and solve equations in ordinary algebra, using plus, times, and minus, so too in Boole's algebra we can write down and solve equations using *or, and,* and *not*.

Even in Boole's time, it was obvious that it should be possible to build machines that carry out Boolean computations, just as we can build machines that perform arithmetic. Such a machine would be a "logic machine," carrying out a piece of logical reasoning. The first example of such a mechanical reasoner was designed and built by an English contemporary of Boole, William S. Jevons. Not surprisingly, Jevons's logic machine was closely fashioned after the calculating machines of the time, namely, mechanical cash registers.

The first transistor, assembled by inventors at Bell Laboratories in 1947. Transistors revolutionized the electronics industry, paving the way for today's microprocessors.

While Jevons's mechanical reasoner showed that it was in principle possible to build a logic machine, it was not until the late 1940s that the technology was available to build really useful machines that could carry out compu-

A map of the inside of a Pentium processor chip shows how far the electronics of computers has come.

tations in Boole's algebra—machines we now call computers.

The basic components of a modern computer are electronic switches called "logic gates" that operate in the same way as the Boolean operations OR, AND, and NOT. They are called "OR gates," "AND gates," and "NOT gates," respectively. (To be perfectly honest, because of engineering considerations, the gates in most present-day computers are not quite of this form, but the idea is essentially the same. Earlier computers did use exactly the approach described here.)

In the case of an OR gate, it has two incoming wires (nowadays, two incoming channels in a silicon chip) and one outgoing wire (or channel). Current flowing in each of the two incoming channels means that the thought represented by that channel is true. The OR gate works by allowing current to flow out whenever at least one of the two input channels carries current. This means that the output channel will carry current precisely if the OR combination of the two input thoughts is true.

The Pentium chip as it appears in its casing.

The AND gate is set up similarly, except that current flows out precisely when current flows in along both input channels simultaneously. For the NOT gate, there is just one input channel and one output channel, and current flows out precisely when current does not flow in.

By joining together large numbers of logic gates in the right way, we can build computers to perform a wide variety of reasoning tasks. How do engineers know how to join the gates together? Boole's mathematics tells them how, by providing the basic patterns of reasoning. The flow of current through the computer's circuits, from gate to gate, obeys the algebraic rules discovered by Boole. Since those rules are rules of thought, that means that the computer is, in a certain sense, "thinking."

"As we move into the era of an information-based world, the value of mathematics as a former of our culture is going to become greater and greater."
KEVIN KELLY
writer and publisher

The historical developments and technical ideas just described are undoubtedly what Media Lab computer scientist Bruce Blumberg had in mind when he said, "Without mathematics, you wouldn't have computers." Mathematics—in particular, Boole's algebra of thought—is the soul of every computer.

These days, computers really are everywhere. This running shoe has been equipped with sensors that measure the impact of various parts of the shoe with the ground. The sensors allow the jogger's wearable computer to gauge the wearer's pace so that, for example, the jogger can pace himself or herself against a distant jogging partner.

These days, computers are everywhere, not just on our desks. Computers are hidden in our televisions and stereos, in our cars, in our wristwatches, and in our microwave ovens and other household appliances. Computers make our telephone connections. Computers are used to make our airline and hotel reservations when we make a business trip or go on vacation. Computers perform many of the tasks involved in piloting a modern airliner. Increasingly, many of the images we see on the cinema screen are produced on a computer. The entire television series *Life by the Numbers* was edited on a computer. This book was written on one computer and typeset on another.

We are, indeed, living in a mathematical universe. Kevin Kelly, a writer and the publisher of the monthly magazine *Wired,* puts it this way: "I think people believe that mathematics has nothing to do with them. This is because mathematics has succeeded in becoming largely invisible, at the same time it has become essential to our lives. As we move into the era of an information-based world, a symbolic-based world, the value of mathematics as a former of our culture is going to become greater and greater, even though it becomes at the same time more and more invisible."

THE SHRINKING WORLD

Like Kevin Kelly, telecommunications engineer Bill Massey sees the world we live in as a world created using mathematics. From the main AT&T control room in New Jersey, Massey has a privileged overview of a rapidly shrinking world—a world made increasingly smaller by the growing network of telephone connections that link us together.

The first articulate sentence ever spoken over an electric telephone was uttered over the instrument below on March 10, 1876. The historic words were "Mr. Watson, come here; I want you," uttered by Alexander Graham Bell after he spilled on his clothes some of the acid that was part of the transmission apparatus. At right is the receiver.

Mathematics is everywhere in telecommunications. The moment you pick up a telephone to call a friend, you are entering a mathematical world. But not because of the billing system. Though that is often what many people think of when trying to think of how mathematics might be used on the telephone system, there is very little mathematics involved in calculating the cost of your call. It just requires keeping track of the time and distance of the connection and charging accordingly: a simple arithmetical calculation, that's all. The real mathematics is elsewhere in the system.

One place where you find mathematics is

Almost twenty years after the first telephone transmission, Alexander Graham Bell places the first New York–to-Chicago call in 1892.

in the software that figures out how to route your call: which path should the system choose from the vast network available? In today's phone system, the decision takes account of the amount of traffic currently on each part of the network, so that the overall load is spread as evenly as possible across the entire network. Finding the best route is a significant challenge. In fact, there is no known method for finding the absolutely best route—there are so many possibilities that even the most powerful computers could not compare them all.

To get some idea of just how many possible paths there are, suppose you had twelve cities, each connected to each of the others by a road. How many different ways are there of visiting all twelve cities,

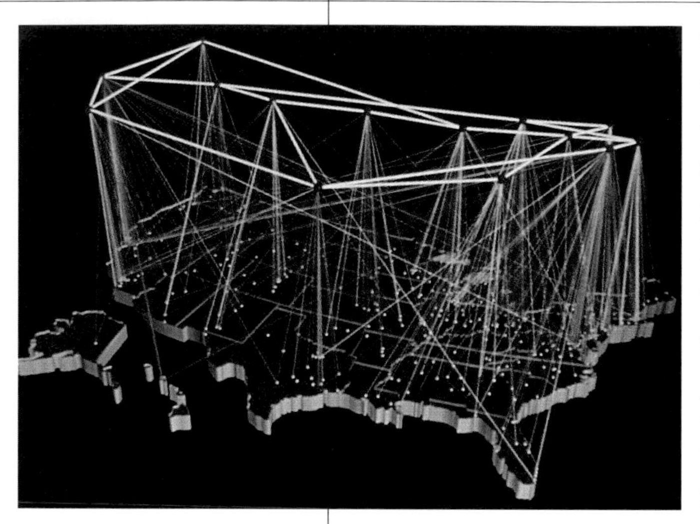

The main network of telephone linkages in the United States.

assuming you visit each city exactly once? The answer is staggering: 479,001,600. That's right, there are almost half a billion different ways of making a tour of just twelve cities. How could you find the most efficient tour, the one that takes the least amount of time?

Now add one more city, to make thirteen in all. How many tours are there now? The answer is 6,227,020,800, over 6 billion. For fifteen cities there are over 1,300 billion tours.

Now try to imagine finding an efficient path of connections through a telephone network (= "road system") with thousands of connecting points (= "cities").

With the number of possible paths being astronomically large, and there being no known method of finding the most efficient path, today's phone system simply tries to find an acceptably good path. Because of the vast number of possibilities, even doing that requires some very sophisticated mathematics, and mathematicians are continually trying to improve the method. All of the methods used require advanced mathematics developed within the past fifty years. One of the best methods known uses ideas from the geometry of spaces with thousands of dimensions. It is so efficient that it takes just a fraction of a second of your having completed dialing in order to complete the computation of the chosen route.

Finding an efficient path through the phone network is just one of a number of mathematical problems involved in making the phone system work. Designing the network in the first place is another. That's where Bill Massey comes in.

Massey works in a branch of mathematics called queuing theory—*queue* being the British word for a waiting line. Queuing theory studies the patterns that arise when queues (or waiting lines) form: queues of cars in heavy traffic, lines of people waiting for a bus, production lines, the flow of current through power lines. Queuing theory can explain that curious phenomenon when traffic on a freeway alternates between stretches of fast driving and periods when everyone slows down and the cars bunch up, even though there seems to be no physical obstruction that would cause the slowdowns.

Though it has many applications, queuing theory was originally developed to understand problems in the telecommunications industry—the application Bill Massey is interested in. He uses queuing theory to try to ensure that calls are connected quickly.

In the United States, each day more than 200 million calls are made. Each one is connected in under three seconds. In the early days of the phone system, when calls were connected by a human operator, it took much longer. The increase in speed between then and now is due in large part to automation. But that isn't the whole story. For automation was accompanied by an enormous growth in the size of the telephone network. What saved the system from being constantly overloaded was queuing theory.

Bill Massey in the AT&T control room.

Using queuing theory, telephone engineers such as Massey can figure out how many trunk lines—the main connections between local exchanges—are needed to ensure that calls are connected quickly. When there are too few trunk lines, callers encounter delays during busy periods. Too many trunk lines, and the system is inefficient. Getting it right requires some sophisticated mathematics. As Massey says, "The more interconnectedness you achieve as individuals, the more coordination you have to have to make the system work properly. That's where science, technology, and, ultimately, mathematics becomes very critical."

"Maybe people don't realize that mathematics is not dead. There are a lot of unsolved problems in mathematics. And there are a lot of things about the world we don't know, and it seems that maybe the only way we'll ever know them is through the application of mathematics."

NATE DEAN
data miner

THE DATA MINER

Nate Dean is a karate expert, a keen photographer, a mathematician, and an explorer—a very new breed of explorer. The unknown terrain that Dean the explorer investigates is the world of data. He looks for hidden patterns among masses of data. He is what is known as a data miner.

"People store data in lots of places—supermarkets, banks, companies, schools, libraries, all kinds of industries. These organizations collect information about people or events. In data mining, we try to extract essential details or information from that data."

The branch of mathematics Dean uses to mine for information is called graph theory. The name is a bit misleading. To most people, a graph is a familiar *x, y* data plot. But mathematicians have another kind of graph, and it is this other kind of graph that is the focus of attention in graph theory. When Dean uses the word *graph,* he means a collection of points drawn on a sheet of paper (or a computer screen), with some of them connected to others by lines. The points are called "nodes" of the graph; the connecting lines are called "edges." Graphs of this kind are also called "networks."

Nate Dean, karate enthusiast and data miner.

Dean is a member of the technical staff at AT&T Bell Laboratories. Much of his work is secret, involving tracking telephone fraud and theft. But he can demonstrate the power of graph theory for the data miner using something as common as grocery store receipts.

Taking a stack of receipts, Dean enters into the computer each item bought by one person over a period. Each item is displayed on the computer screen as a dot. The dots are the nodes of the graph Dean draws on his computer. He connects two dots by a line (an edge of the graph) if the corresponding two items were bought at the same time. For example, if the customer bought soap, dishwashing deter-

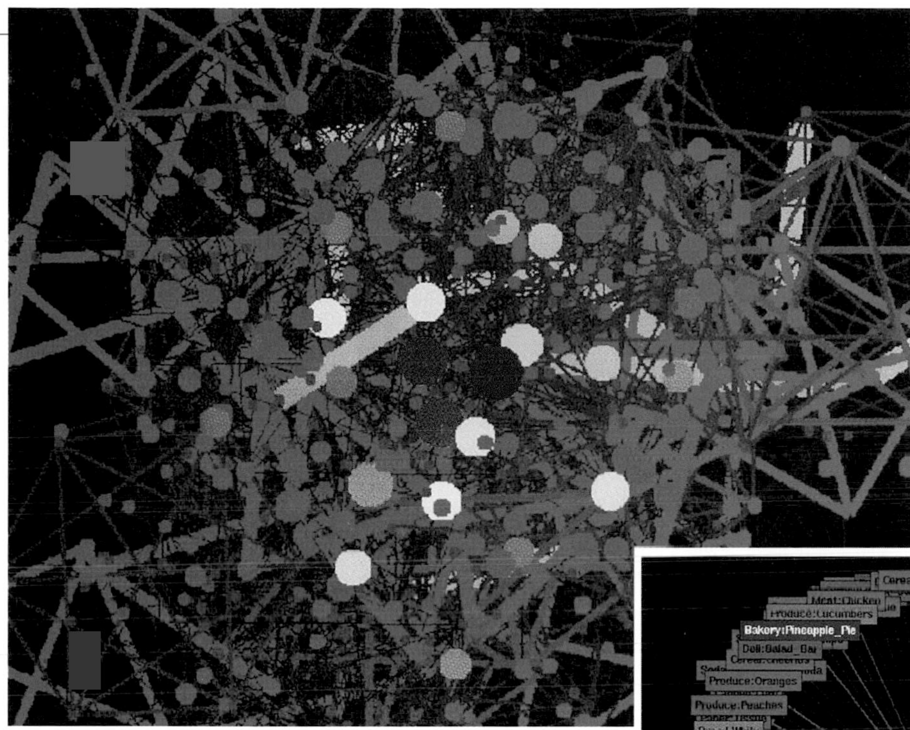

This is the image Nate Dean creates by entering all of his data about product purchases into his computer. The red dot at the center represents the highly interlinked item of bananas.

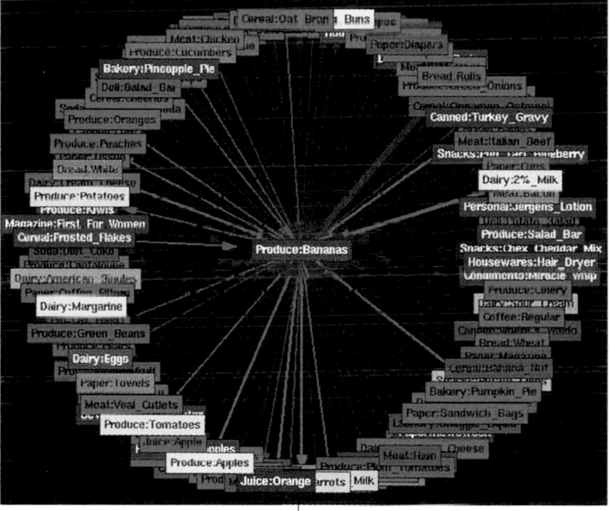

This alternate representation of the graph lists the products by name and much more clearly shows the central position of bananas in the purchasing history.

gent, and potatoes at the same time, the three items' dots would all be connected to each other by edges.

Starting with a number of receipts, the graph Dean obtains in this way can soon look pretty complicated. To try to make sense of it, he alters it, moving some nodes toward the center and others out away from the center. He does this to try to reveal those items purchased most frequently. The resulting graph reveals one particular item that is connected to just about every other item: bananas. This customer bought bananas on almost every visit to the store. Dean has discovered a purchasing pattern.

"There certainly is a lot of data mining going on now," Dean remarks. "Everybody seems to be jumping into it. It's a hot topic.

We find that if we look at data the right way, the same patterns start to emerge. Maybe these patterns don't explain everything, but we start to understand a little bit about why people are buying things the way they do. We start to understand things about human behavior and about ourselves."

THE PULSE OF AMERICA

Nate Dean uses graph theory to find hidden patterns in data. Walker Smith also looks for patterns in data. But the mathematics he uses is statistics. As a partner in Yankelovich Partners, a strategic market research firm, Smith uses mathematics to keep his finger firmly on the pulse of America.

"It's mathematics that has really enabled us to have the tools and capabilities of understanding what the American Dream is, and of tracking it over time. And it's mathematics that's going to enable us to keep up with it in the future."

WALKER SMITH
market researcher

"We work with companies that sell products and services to help them understand the values and the buying motivations of their customers," Smith explains. "We do primarily survey research—we interview customers—what I would call research the old-fashioned way."

Such is the power of modern statistics that, by interviewing a thousand carefully selected people, asking questions that have been formulated with enormous precision, Walker Smith and his associates can predict the behavior of 240 million Americans to an amazing degree of accuracy, often within one or two percentage points.

For over fifty years, Yankelovich Partners has been using statistical mathematics to track the American Dream, and the firm was often among the first to spot a new trend. For example:

- In the 1960s, 43 percent of college-educated youth held sexual freedom as a "very important value." And the trend was rising.
- In the 1970s, 42 percent of Americans were actively looking for new foods to eat at home. And the trend was rising.

- In the 1980s, 24 percent of Americans believed that money was the only meaningful measure of success. And the trend was rising.

- Today, 71 percent of Americans believe that life has become much too complicated, and 74 percent believe that technology is so confusing that it is hard to know what brands or models of a product to buy. And the trends are rising.

Once the door-to-door researchers have collected the interview data, all the data are entered into the computer, where they become rows of numbers. The computer starts to work on the numbers. But, as Smith observes, the numbers are just a way of getting to the answers: "The data we collect are more than just a bunch of numbers. The numbers that we process and analyze really correspond to opinions that people express." For example:

- Opinion: 80 percent of Americans are looking for ways to simplify their lives.

- Opinion: Fewer than 40 percent of Americans feel the need to keep up with new styles.

- Opinion: Fewer than 20 percent of Americans buy brands to let other people know they've made it.

"The number one sign of success and accomplishment used to be traveling for pleasure," Smith observes. "The number one sign of success and accomplishment nowadays is being satisfied with your life and being in control."

- Number of Americans satisfied with their lives: 79 percent and rising.

- Number of Americans in control of their lives: 78 percent and rising.

- Number of Americans who have a good marriage: 77 percent and rising.

"We have seen the American Dream go through some radical transformations," Smith comments. "It's going to change a lot again in the future, and it's mathematics that has really enabled us to have the tools and capabilities of understanding what the American Dream is, and of tracking it over time. And it's mathematics that's going to enable us to keep up with it in the future."

HEADING INTO THE SHIFT

Changing trends are also what interests Graciela Chichilnisky. Whereas Walker Smith tracks changes within American society, Chichilnisky looks at the global economy. One of the world's leading economists, she studies the changes that the world's economies are undergoing, and tries to predict what will happen in the future. According to her analysis, the world economy is currently evolving in some very profound ways.

"We are facing a major economic change," she says. "The world

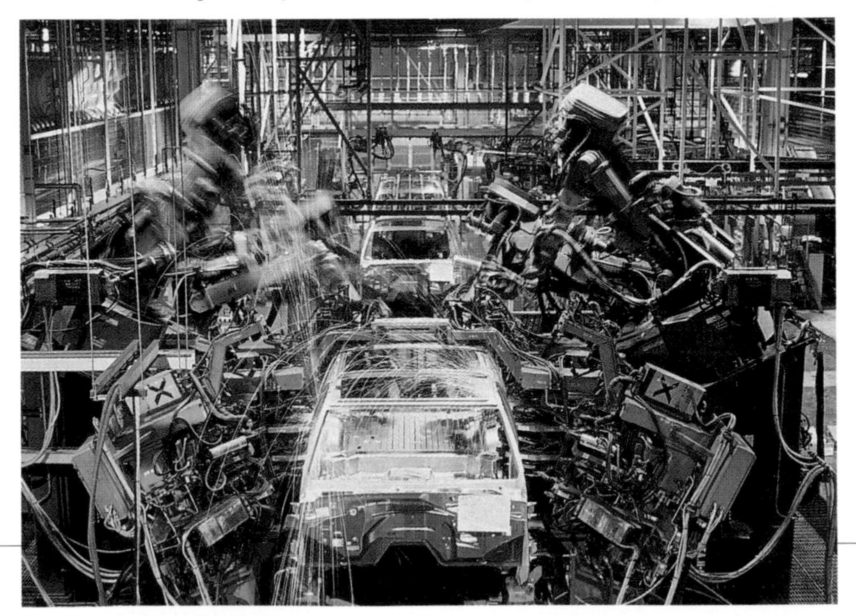

This automobile assembly line is operated entirely by robotic machines, which have replaced human workers in many areas of industrial production.

economy is about to do a somersault. In the face of that, we are changing the way we produce goods, the way they circulate, the way we work, the way we think about things. We are now seeing a new wave of change."

Chichilnisky sees us as entering a new, fourth era of human history. In the first era, she explains, humans were hunters and gatherers, relying on physical strength and their ability to find food. In the second era, humans became farmers, relying on the land. The third era was the industrial age; people relied on fossil fuels and machinery.

We are just now entering the fourth era, says Chichilnisky. "Now, industrial society is becoming a different type of society. This different type of society I call the knowledge society. And the transition we are going through I call the knowledge revolution."

Many people might be surprised at what Chichilnisky thinks is the fuel driving this change. "Mathematics is the raw material, it's the energy that drives the system," she suggests. "Mathematics works for today's society like the fossil fuels worked for industrial society. Today, to get energy, we don't burn fossil fuels. Now, to get knowledge, we burn mathematics, we use mathematics."

For Chichilnisky, mathematics is also the key tool to understanding the changes starting to take place: "In addition to being the driving force between these changes, mathematics is very important for our ability to understand these changes and being able to use this understanding—to formalize it, to conceptualize it, and to see where we're going, what's happening, and why."

Chichilnisky uses a branch of mathematics called topology to visualize the growth of national economies. Topology studies very general properties of figures, properties that do not depend on their actual shape and size. For example, being a "ball" is a topological

"Mathematics is at the core of the most exciting applications today. All of the very dynamic sunrise sectors [of the world's economies] are merging between themselves and altering themselves through the use of mathematical tools."

GRACIELA CHICHILNISKY
economist

property, since balls come in many different shapes and sizes. For the topologist, all balls are the same.

In topology there are no precise measurements, just general properties of shape: whether something is "ball-shaped" or "ring-shaped" or "pretzel-shaped" or whatever. It is sometimes referred to as "rubberland geometry" to indicate that it can be regarded as the study of properties of objects made of a perfectly elastic material. In topology, a golf ball is "equal to" a football (American football or soccer—it makes no difference in topology), and to a large beach ball. But no ball is "equal to" a rubber ring or a life belt, since you cannot deform a ball into a ring.

It is the absence of measurement that made topology ideally suited to Chichilnisky's work. "In economics we are dealing with large structures," she explains. "It's very difficult to predict any single thing that will happen, but large shapes we can predict. In my own work I have introduced the study of certain cones, and the topology of these cones, to understand how markets operate."

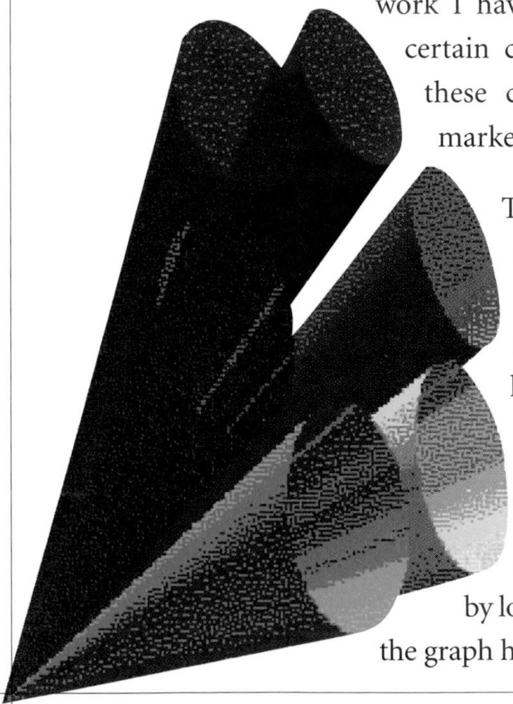

These production cones represent six different economies plotted on three dimensions. The vertical dimension represents the amount of goods produced, the horizontal dimension represents the amount of knowledge used to produce the goods, and the third dimension—which we can imagine as receding back from the surface of this book page—represents the amount of natural resources used to produce the goods. The most knowledge-intensive economy—and the one best equipped to move into the new knowledge economy—is the one represented by the green cone.

The "cones" Chichilnisky is referring to are a special kind of graph she invented to represent economies. Usually her graphs have hundreds or even thousands of dimensions, so they can't be drawn. But you can understand how her graphs work by looking at a simple case, where the graph has just three dimensions.

For instance, Chichilnisky uses her cones to examine production—in this case, in three dimensions. One axis of the graph represents the quantity of goods produced by a country. Zero denotes no goods produced. As we move along the axis, more and more goods are made. The second axis represents the amount of natural resources required to produce the goods. Zero denotes no resources used; as we move along the axis, more resources are used. The third axis represents the amount of knowledge used to produce the goods. As we move along the axis, more knowledge is required.

When Chichilnisky plots a country's economic activity on such a graph, it comes out to be in the shape of a cone—a production cone. The shape of the cone tells her a lot about the efficiency of the economy. For example, a cone that is fat in the knowledge direction tells us that the economy uses its knowledge very effectively. A cone that is thin in both the resources and knowledge directions makes inefficient use of both resources and knowledge.

By taking the data from the world's national economies and plotting them as cones on her graph, she has observed a simple relationship. For a given amount of natural resources, the more knowledge an economy uses, the more productive it will be. Chichilnisky sees this observation as having significant implications for the future.

Says Chichilnisky: "Knowledge has always been important for anything we do. But the movers and shakers always felt contempt for people who traded in knowledge. You have probably heard the sentence 'If you are so smart, why aren't you rich?' Well, the situation has changed. The richest ones are the smartest ones. And it is understood now that it is knowledge which is attracting capital."

The growing importance of knowledge, highlighted so dramatically in Chichilnisky's cones, is starting to have a major effect not just on individual economies but on the overall nature of world trade.

Using another set of cones, called "market cones," Chichilnisky has developed another theory, one that many will find unsettling.

An economy that is heavily reliant on knowledge, such as the U.S. economy, will have a narrow market cone. The range of potential prices within the economy is small, and there is not much fluctuation in the stock market. An economy that relies more on natural resources will have a broad market cone, which means that the potential range of prices is large. Mexico is an example of such an economy.

By drawing the market cones for the different nations in the world, Chichilnisky can see how world trade is likely to develop. For example, the intersection of two cones marks the spot where prices are the same and trading can take place. Cones that don't touch represent countries that don't trade.

When Chichilnisky puts together the results of her research, her cone graphs show in a very dramatic fashion that the world's economies are separating into two distinct classes. The knowledge-intensive economies will increasingly just trade among themselves, leaving out the resource-based economies. Chichilnisky believes that many of the modern trading alliances and blocks we see forming today are a result of this relationship.

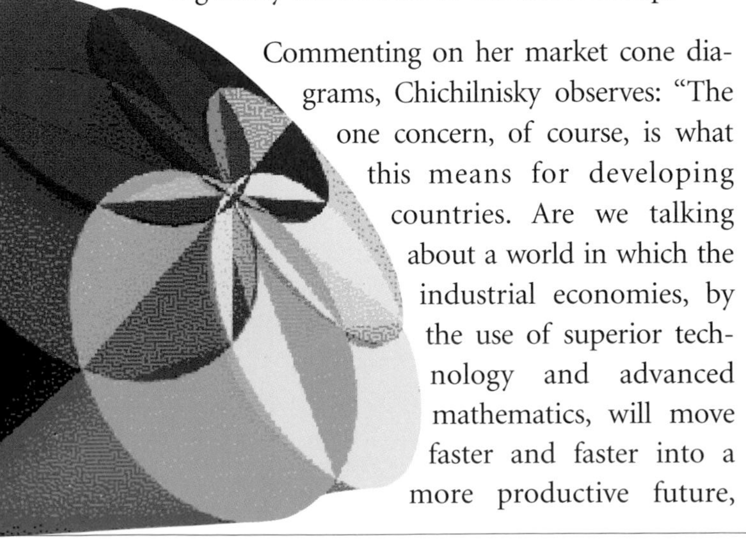

These market cones represent six economies that trade heavily with one another—such as might be seen for the economies of Western Europe—which is represented by the way the cones overlap extensively.

Commenting on her market cone diagrams, Chichilnisky observes: "The one concern, of course, is what this means for developing countries. Are we talking about a world in which the industrial economies, by the use of superior technology and advanced mathematics, will move faster and faster into a more productive future, leaving the developing countries and the agricultural society further and further behind?"

Chichilnisky says that she cannot predict the future. The future of society is in our hands, she says. Mathematics can often provide models that lead to useful hypotheses, and in the inanimate world of physics and chemistry, the predictions of mathematics are generally remarkably accurate. In the animate world of people and economies, however, things are different, because people can alter the future. Chichilnisky's cones make a prediction about what might happen. It is a prediction that appears to be borne out by what is happening already. But only time will tell for sure. Of course, by taking seriously the mathematical models produced by researchers such as Chichilnisky, it may be possible to introduce changes to redirect or alleviate the way things develop.

"Mathematics works for today's society like the fossil fuels worked for industrial society."
GRACIELA CHICHILNISKY
world economist

Says Chichilnisky: "So we are living through the shift, through this change. It's a bit scary. Although we know something is changing, we don't know where we're going. Nobody knows where we are going. We haven't been there before."

Graciela Chichilnisky with the floor of the New York Stock Exchange in the background.

People use mathematics to help see the invisible, to try to understand what would otherwise not be seen: from the origins and the far reaches of the universe to the bottom of the ocean; from the patterns of chance events to the internal workings of the human mind. Chichilnisky uses mathematics to see into the future. Her simple cone diagrams are both visually attractive and profoundly important.

"Many people believe that mathematics is boring and complicated," Chichilnisky declares. "But in fact these pictures of cones show that mathematics can be beautiful and simple. Some of the most interesting phenomena we are trying to explain can be captured with our geometric intuition, simply by observing these beautiful pictures."

Chapter 8

IT'S AN
M WORLD

The ancient Greek philosopher Plato thought that mathematics was the supreme form of knowledge and the key to all other knowledge. To enter Plato's Academy, you had to know mathematics. For the Greeks, knowledge of mathematics was part of what it took to be regarded as an educated and cultured person.

Today's technological world is a direct descendant of the intellectual and cultural tradition begun by the Greeks. It is a world in which mathematics plays a far greater role than in Plato's time. Indeed, much of our present world is the product of mathematics. And yet, today, knowledge of mathematics is not generally regarded as important—a person can be largely ignorant of mathematics and still be regarded as educated.

Dozens of explanations have been given for the change in attitudes toward mathematics. This is not the place to discuss them. But the most significant is almost certainly that, over the years, as mathematics became more and more complicated, people concentrated more and more on the numbers, the formulas, the equations, and the methods, and lost sight of what those numbers, formulas, and equations were really about and why those methods were developed. They lost sight of the fact that mathematics is not about manipulating symbols according to arcane rules but is about understanding patterns—patterns of nature, patterns of life, patterns of beauty.

To take just one example, think of Graciela Chichilnisky's use of mathematics in global economics, as described at the end of the previous chapter. Beautiful, imaginative, creative, simple, important, relevant—all of these adjectives apply to her work. And yet, for many people they are not words normally associated with mathematics. Most people think that mathematics makes the world more complicated. It doesn't do that.

When you get beyond the symbols and the formulas, you find that mathematics actually makes the world more simple. Mathematicians strip away the complexity and look at the world in the simplest possible way. They look at it in such a simple way that the only way of capturing that simplicity is with symbols, with numbers, with algebra, and with graphs.

That simplicity is what gives mathematics its incredible power—power to help us understand, power to help us do good, and, if we are not wary, power to do harm.

With the power and the simplicity of mathematics, we can do many things. We can create imaginary worlds, on the movie screen or in the computer, we can investigate the strange world of the fourth dimension, and we can look back in time to discover how dinosaurs must have moved. We can explain how leopards get their spots, investigate how viruses attack the body, and find ways to improve the performances of athletes. We can draw maps to help us find our way around, and we can design and build machines that perform calculations for us, and that perhaps even one day can "think." Mathematics does all this for us and more. It does so by providing us with a way to *make the invisible visible.*

Newton's equations of motion, discovered in the seventeenth century, let us "see" the invisible forces that keep the earth rotating around the sun and cause an apple to fall from the tree onto the ground. An equation discovered by the mathematician Daniel Bernoulli early in the eighteenth century allows us to "see" the invisible force that keeps an airplane in the air. Two thousand years before we could send spacecraft into outer space to provide us with pictures of our planet, the Greek mathematician Eratosthenes used mathematics to show that the earth is round, calculating its diameter with 99 percent accuracy.

Using mathematics and powerful telescopes, we can "see" into the outer reaches of the universe, and perhaps one day soon we will discover the shape of the universe. We can use mathematics to see back to the otherwise invisible moments when the universe was first created in the Big Bang. Probability theory enables us to look into the future, to predict the outcomes of elections, often with remarkable accuracy. Insurance companies use statistics and probability theory to predict the likelihood of an accident during the coming year, and set their premiums accordingly. We can even use calculus to predict tomorrow's weather.

The list goes on. Sometimes, even though we are surprised by a particular application of mathematics, we readily accept it because it is in an area where we expect to be able to use mathematics. Many applications of mathematics in physics and the other sciences are like that. Occasionally, however, we are surprised, when we meet a use of mathematics in an area that we don't think of as being precise or "mathematical." For example, when we discover that mathematics can be used to "see" the invisible forces of nature that give a flower its intricate and beautiful shape. Or when we learn that Aristotle used mathematics to try to "see" the invisible patterns of sound that we recognize as music and to "see" the invisible structure of a dramatic performance.

In the 1950s, the linguist Noam Chomsky surprised everyone by using mathematics to "see" and describe the invisible, abstract patterns of words that we recognize as a grammatical sentence. Blaise Pascal, who with Pierre de Fermat invented probability theory, even used probability theory to try to prove to people that they should live a pious life, arguing that no matter how small the probability is that God exists, when you multiply that probability by the infinite payoff of eternity spent in heaven, the result is an infinite "expectation" that far outweighs the finite expectation you get if God does not exist.

Whether or not we take Pascal's advice when deciding how to live our lives, these days we live those lives in a mathematical universe. The house we live in and the automobile we use to get around were both designed using mathematics. So too was the airplane we fly in—and these days planes fly and navigate by means of mathematics as well. Our hospitals are full of equipment designed using mathematics, and the medicines we receive are tested using mathematics. Mathematics lies behind the telephone system, behind television and radio, and behind CD players. Mathematics is used to decide what products are available for us to buy in the store and what programs are available to watch on television. Computers—another product of mathematics—are everywhere and affect many aspects of our lives. Movies are often made using mathematical techniques. Mathematics is playing an increasing role in sports and recreation activities. We can use mathematics to take the ideas and thoughts produced by our imaginations and make them accessible for others to share. The list could go on, but the message is clear: Today's world is in large part a world of mathematics.

Mathematics is a product—a discovery—of the human mind. It enables us to see the incredible, simple, elegant, beautiful, ordered structure that lies beneath the universe we live in. It is one of the greatest creations of mankind—if it is not indeed the greatest.

Further Reading

If you'd like to discover more about the world of mathematics, try any of the following books. All are written for the general reader. They are listed here in what I judge to be increasing order of difficulty.

Peterson, Ivars. *The Mathematical Tourist: Snapshots of Modern Mathematics.* W. H. Freeman, 1988.

———. *Islands of Truth: A Mathematical Mystery Cruise.* W. H. Freeman, 1988.

Bernstein, Peter L. *Against the Gods: The Remarkable Story of Risk.* Wiley, 1996.

Stein, Sherman. *Strength in Numbers: Discovering the Joy and Power of Mathematics in Everyday Life.* Wiley, 1996.

King, Jerry P. *The Art of Mathematics.* Plenum, 1992.

Stewart, Ian. *Nature's Numbers: The Unreal Reality of Mathematics.* Basic Books, 1996.

Schattschneider, Doris. *M. C. Escher: Visions of Symmetry.* W. H. Freeman, 1990.

Devlin, Keith. *Mathematics: The Science of Patterns.* W. H. Freeman, Scientific American Library series, 1994, 1996.

Dunham, William. *Journey Through Genius: The Great Theorems of Mathematics.* Penguin, 1991.

Devlin, Keith. *Mathematics: The New Golden Age.* Penguin, 1987, 1998.

Dunham, William. *The Mathematical Universe: An Alphabetical Journey through the Great Proofs, Problems, and Personalities.* Wiley, 1994.

Stewart, Ian. *The Problems of Mathematics.* Oxford University Press, 1992.

To learn more about the following subjects, you are invited to visit the Web sites listed:

The Koch Snowflake and fractals. Mary Ann Connors's Web site, Exploring Fractals, at http://www.math.umass.edu/~mconnors/fractal/fractal.html

Mathematical knots. Rob Scharein's KnotPlot Site, at http://www.cs.ubc.ca/nest/imager/contributions/scharein/KnotPlot.html

Cartography. The U.S. Geological Survey's site about mapping at http://mapping.usgs.gov

Credits

(T) = Top; (B) = Bottom; (M) = Middle; (L) = Left; (R) = Right

Front Matter

Title Page: left side: Robert Frerck/Tony Stone Images; right side: (T) Tim Davis/Tony Stone Images, (M) Corbis-Bettmann, (B) WQED Pittsburgh

The Invisible Universe

1, 2 NASA; 3 Renee Lynn/Tony Stone Images; 5 Shelby Lyons; 6 Robert Williams and the Hubble Deep Space Field Team (STScl) and NASA; 7 University of Illinois, Urbana-Champaign, artists Donna Cox and Robert Patterson, simulation of galaxies by Professor Susan Lamb and Richard Gerber, University of Illinois, Urbana-Champaign; 9 Dawn Wright; 10(L) Oliver Meckes/Gelderblom/Photo Researchers, Inc.; 10(R) LSHTM/Tony Stone Images; 11 Sylvia Spengler; 12(L) Gary Benson/Tony Stone Images; 12(R) Art Wolfe/Tony Stone Images; 13 Clifford Pickover; 14 Trinity College Library, Cambridge University; 15 The Pierpont Morgan Library; 17 Erich Schrempp/Photo Researchers, Inc.; 18 Dr. Jeremy Burgess/Science Photo Library

Seeing Is Believing

21 The Granger Collection; 22–23 Photofest; 24, 25 Entertainment Design Workshop, LLC; 26 Archive Photos/David Lees; 27 The Granger Collection; 28 Scala/Art Resource; 29 Girandon/Art Resource; 30 WQED Pittsburgh; 31 University of Illinois, Urbana-Champaign, photo by Bill Wiegand; 32 Photofest; 33, 34, 36 WQED Pittsburgh; 37 General Research Division, New York Public Library; 38 Jonathan Bowen, University of Reading; 40(T) The Granger Collection; 40(B) The Museum of Modern Art, New York. André Meyer Bequest. Photograph © 1997 The Museum of Modern Art, New York; 41 WQED Pittsburgh; 42 Tony Robbin; 44 WQED Pittsburgh; 45, 46 Marcos Novak

Patterns of Nature

49 Renee Lynn/Tony Stone Images; 50 Corbis-Bettmann; 52 Photofest; 53 WQED Pittsburgh; 54(T) WQED Pittsburgh; 54 55 Kingston Museum & Heritage Service, Surrey, England; 56 Photo by Denis Finnin/American Museum of Natural History; 57 Painting by Robert J. Barber, photo by Denis Finnin/American Museum of Natural History; 58 Renee Lynn/Tony Stone Images; 59 Tim Davis/Tony Stone Images; 60–61, 63 WQED Pittsburgh; 65–67 Knot images by Rob Scharein, Imager Computer Graphics Laboratory, University of British Columbia, using the KnotPlot software; 69 WQED Pittsburgh; 70 Sylvia Spengler; 72 Briggite Merle/Tony Stone Images; 73(T), 73(M) NASA; 73(B) The Geometry Center, University of Minnesota (Clouds 2 by Danek Duvall); 74 Images copyright of Mary Ann Connors, used by permission; 75(T) IBM Thomas J. Watson Research Center; 75(B) Clifford Pickover; 76 Dr. Campbell Davidson and Dr. Przemyslaw Prusinkiewicz; 79 Robert Frerck/Tony Stone Images; 82 Courtesy of Tom Ray

The Numbers Game

85 The Harold E. Edgerton 1992 Trust, courtesy of Palm Press, Inc.; 86 Villa of Casale, Pizza Armerina, Sicily/Silvio Fiore/Superstock; 87(L) Columbia University; 87(R) Columbia University; 88 Wiley Photo Library; 89 Dr. Gary Settles/Science Source; 90, 91 Rebus, Inc.; 92 Otto Creule/Allsport USA; 93 The Harold E. Edgerton 1992 Trust, courtesy of Palm Press, Inc.; 95 John Gonzalez; 97 Used by permission from the New York Yacht Club, Mystic Seaport Museum, Inc., Rosenfeld Collection; 98 Bob Greiser; 101 (T) Boeing; 101(M) South Bay Simulations, Inc. The Pressure contours of an International America's Cup Class yacht are identified with SPLASH free-surface code. 101(B) Young America; 102, 103(L), 103(R) WQED Pittsburgh; 105 G. Brad Lewis/Tony Stone Images; 106, 107, 109 WQED Pittsburgh; 110, 111, 112, 113 Amy Jo Haufler

The Shape of the World

117 United States Geological Survey; 118 The Granger Collection, New York; 120 Nigel Holmes; 121 *Mercator Atlas,* New York Public Library Special Collections; 122, 123 Charles Ehlschlager; 125 SPOT Image. This perspective was produced by the Defense Mapping Agency, now the National Imagery and Mapping Agency, in support of the Bosnia Peace Talks held in Dayton, Ohio. It is derived from SPOT imagery, SPOT Image Corporation, Reston, VA; 127, 128, 130, 131 WQED Pittsburgh; 132(T), 132(B), 133 Dawn Wright; 134 WQED Pittsburgh; 135 Tony Craddock/Science Photo Library; 136 David Parker/Science Photo Library; 139 Lynette Cook/Science Photo Library; 140 Chun Shing Jason Pun (NASA/GSFC), Robert P. Kirshner (Harvard-Smithsonian Center for Astrophysics), and NASA; 141 NASA; 142, 143, 144, 145 Jeff Weeks

Chances of a Lifetime

147 John Warden/Tony Stone Images; 148–149 John Warden/Tony Stone Images; 151 Arte romana: Giuocatrici d'astragali prov. da Ercolano, Napoli, Museo Nazionale/Art Resource; 152 Luigi Sabatelli. Olympus. Sala dell'Iliade. Ceiling fresco. Galleria Palatina, Palazzo Pitti, Florence, Italy; Scala/Art Resource; 153 Corbis-Bettman; 154 IBM Corporation; 155(T) Columbia University; 155(B) PhotoDisc; 156 Andy Sacks/Tony Stone Images; 158 Stephen Dunn/Allsport USA; 161 UPI/Corbis-Bettmann; 163 Black Star; 164 New York Public Library Special Collections; 166 River Graphics/The Gallup Monthly Poll; 169 Lloyd's of London; 170–171 Steve Bronstein/The Image Bank; 171 (inset) Chris Johns/Tony Stone Images; 173 PhotoDisc; 175 NASA; 176 David J. Roddy/United States Geological Survey, Flagstaff, AZ

A New Age

179 Webb Chappell; 180, 181 WQED Pittsburgh; 182, 183, 184(L), 184(R) MIT Media Lab; 185 Science Photo Library; 186 IBM Archives; 187 Bell Labs; 188(T), 188(B) Intel Corp.; 189 Steve Mann; 190 AT&T Photo Center; 191 AT&T Archives; 192 National Center for Supercomputing Applications, University of Illinois, Urbana; 193, 194 WQED Pittsburgh; 195(T), 195(B) Nathaniel Dean, courtesy of Lucent Technologies; 198 Mark Segal/Tony Stone Images; 200, 202 Graciela Chichilnisky; 203 WQED Pittsburgh

It's an M World

205 Tony Craddock/Science Photo Library

With special thanks to the following researchers and organizations, who so kindly offered their assistance in providing images:

Russell Berry of the U.S. Geological Survey; Mary Ann Connors; Donna Cox; Paul Dalton of Lloyd's of London; Nathaniel Dean; Jane Eagleson of Young America; John Gonzalez; Linda Halet of The Geometry Center; Amy Jo Haufler; Nigel Holmes; Debbie King; Benoit Mandelbrot; Steve Mann; Valerie Eames Minard of the MIT Media Lab; the National Aeronautics and Space Administration; Kathleen Neary of the National Imagery and Mapping Agency; Marcos Novak; Clifford Pickover; Przemyslaw Prusinkiewicz; R. V. Ramnath; David Roddy of the U.S. Geological Survey; Rob Scharein; Sylvia Spengler; Douglas Trumbull; Jeff Weeks.

Index